Improving Bridge Design

STEM Road Map for Middle School

Improving Bridge Design

STEM Road Map for Middle School

Edited by Carla C. Johnson, Janet B. Walton, and
Erin Peters-Burton

Arlington, Virginia

Claire Reinburg, Director
Rachel Ledbetter, Managing Editor
Deborah Siegel, Associate Editor
Andrea Silen, Associate Editor
Donna Yudkin, Book Acquisitions Manager

Art and Design
Will Thomas Jr., Director, cover and interior design
Himabindu Bichali, Graphic Designer, interior design

Printing and Production
Catherine Lorrain, Director

National Science Teachers Association
David L. Evans, Executive Director

1840 Wilson Blvd., Arlington, VA 22201
www.nsta.org/store
For customer service inquiries, please call 800-277-5300.

21 20 19 18 4 3 2 1

NSTA is committed to publishing material that promotes the best in inquiry-based science education. However, conditions of actual use may vary, and the safety procedures and practices described in this book are intended to serve only as a guide. Additional precautionary measures may be required. NSTA and the authors do not warrant or represent that the procedures and practices in this book meet any safety code or standard of federal, state, or local regulations. NSTA and the authors disclaim any liability for personal injury or damage to property arising out of or relating to the use of this book, including any of the recommendations, instructions, or materials contained therein.

Library of Congress Cataloging-in-Publication Data
Names: Johnson, Carla C., 1969- editor. | Walton, Janet B., 1968- editor. | Peters-Burton, Erin E., editor.
Title: Improving bridge design, grade 8 : STEM road map for middle school / edited by Carla C. Johnson, Janet B. Walton, and Erin Peters-Burton.
Description: Arlington, VA : National Science Teachers Association, [2018] | Includes bibliographical references and index.
Identifiers: LCCN 2018012163 (print) | LCCN 2018007473 (ebook) | ISBN 9781681404158 (E-book) | ISBN 9781681404141 (print)
Subjects: LCSH: Bridges--United States--Design and construction--Study and teaching (Middle school) | Structural analysis (Engineering)--United States--Study and teaching (Middle school) | Infrastructure (Economics)--United States--Study and teaching (Middle school) | Eighth grade (Education)
Classification: LCC TG300 (print) | LCC TG300 .I45 2018 (ebook) | DDC 624.2/5--dc23
LC record available at *https://lccn.loc.gov/2018012163*

The *Next Generation Science Standards* ("*NGSS*") were developed by twenty-six states, in collaboration with the National Research Council, the National Science Teachers Association and the American Association for the Advancement of Science in a process managed by Achieve, Inc. For more information go to *www.nextgenscience.org*.

CONTENTS

CONTENTS

ABOUT THE EDITORS AND AUTHORS

Dr. Carla C. Johnson is the associate dean for research, engagement, and global partnerships and a professor of science education at Purdue University's College of Education in West Lafayette, Indiana. Dr. Johnson serves as the director of research and evaluation for the Department of Defense–funded Army Educational Outreach Program (AEOP), a global portfolio of STEM education programs, competitions, and apprenticeships. She has been a leader in STEM education for the past decade, serving as the director of STEM Centers, editor of the *School Science and Mathematics* journal, and lead researcher for the evaluation of Tennessee's Race to the Top–funded STEM portfolio. Dr. Johnson has published over 100 articles, books, book chapters, and curriculum books focused on STEM education. She is a former science and social studies teacher and was the recipient of the 2013 Outstanding Science Teacher Educator of the Year award from the Association for Science Teacher Education (ASTE), the 2012 Award for Excellence in Integrating Science and Mathematics from the School Science and Mathematics Association (SSMA), the 2014 award for best paper on Implications of Research for Educational Practice from ASTE, and the 2006 Outstanding Early Career Scholar Award from SSMA. Her research focuses on STEM education policy implementation, effective science teaching, and integrated STEM approaches.

Dr. Janet B. Walton is a research assistant professor and the assistant director of evaluation for AEOP at Purdue University's College of Education. Formerly the STEM workforce program manager for Virginia's Region 2000 and founding director of the Future Focus Foundation, a nonprofit organization dedicated to enhancing the quality of STEM education in the region, she merges her economic development and education backgrounds to develop K–12 curricular materials that integrate real-life issues with sound cross-curricular content. Her research focuses on collaboration between schools and community stakeholders for STEM education and problem- and project-based learning pedagogies. With this research agenda, she works to forge productive relationships between K–12 schools and local business and community stakeholders to bring contextual STEM experiences into the classroom and provide students and educators with innovative resources and curricular materials.

Dr. Erin Peters-Burton is the Donna R. and David E. Sterling endowed professor in science education at George Mason University in Fairfax, Virginia. She uses her experiences from 15 years as an engineer and secondary science, engineering, and mathematics teacher to develop research projects that directly inform classroom practice in science and engineering. Her research agenda is based on the idea that all students should build self-awareness of how they learn science and engineering. She works to help students see themselves as "science-minded" and help teachers create classrooms that support student skills to develop scientific knowledge. To accomplish this, she pursues research projects that investigate ways that students and teachers can use self-regulated learning theory in science and engineering, as well as how inclusive STEM schools can help students succeed. During her tenure as a secondary teacher, she had a National Board Certification in Early Adolescent Science and was an Albert Einstein Distinguished Educator Fellow for NASA. As a researcher, Dr. Peters-Burton has published over 100 articles, books, book chapters, and curriculum books focused on STEM education and educational psychology. She received the Outstanding Science Teacher Educator of the Year award from ASTE in 2016 and a Teacher of Distinction Award and a Scholarly Achievement Award from George Mason University in 2012, and in 2010 she was named University Science Educator of the Year by the Virginia Association of Science Teachers.

Dr. Toni A. Ivey is an associate professor of science education in the College of Education at Oklahoma State University. A former science teacher, Dr. Ivey's research is focused on science and STEM education for students and teachers across K–20.

Dr. Tamara J. Moore is an associate professor of engineering education in the College of Engineering at Purdue University. Dr. Moore's research focuses on defining STEM integration through the use of engineering as the connection and investigating its power for student learning.

Dr. Sue Christian Parsons is an associate professor and the Jacques Munroe Professor in Reading and Literacy Education at Oklahoma State University. A former English language arts teacher, her research focuses on teacher development and teaching and advocating for diverse learners through literature for children and young adults.

Dr. Adrienne Redmond-Sanogo is an associate professor of mathematics education in the College of Education at Oklahoma State University. Dr. Redmond-Sanogo's research is focused on mathematics and STEM education across K–12 and preservice teacher education.

Dr. Toni A. Sondergeld is an associate professor of assessment, research, and statistics in the School of Education at Drexel University in Philadelphia. Dr. Sondergeld's research concentrates on assessment and evaluation in education, with a focus on K–12 STEM.

Dr. Juliana Utley is a professor and the Morsani Chair in Mathematics Education in the College of Education at Oklahoma State University. A former mathematics teacher, Dr. Utley's research is focused on mathematics and STEM education across K–20.

John Weaver is a clinical instructor in the College of Education at Oklahoma State University. A former mathematics teacher, he teaches elementary and secondary mathematics methods courses and is a master teacher for the OSUTeach program.

ACKNOWLEDGMENTS

This module was developed as a part of the STEM Road Map project (Carla C. Johnson, principal investigator). The Purdue University College of Education, General Motors, and other sources provided funding for this project.

See *www.routledge.com/products/9781138804234* for more information about *STEM Road Map: A Framework for Integrated STEM Education.*

THE STEM ROAD MAP

BACKGROUND, THEORY, AND PRACTICE

OVERVIEW OF THE *STEM ROAD MAP CURRICULUM SERIES*

Carla C. Johnson, Erin Peters-Burton, and Tamara J. Moore

The *STEM Road Map Curriculum Series* was conceptualized and developed by a team of STEM educators from across the United States in response to a growing need to infuse real-world learning contexts, delivered through authentic problem-solving pedagogy, into K–12 classrooms. The curriculum series is grounded in integrated STEM, which focuses on the integration of the STEM disciplines—science, technology, engineering, and mathematics—delivered across content areas, incorporating the Framework for 21st Century Learning along with grade-level-appropriate academic standards.

The curriculum series begins in kindergarten, with a five-week instructional sequence that introduces students to the STEM themes and gives them grade-level-appropriate topics and real-world challenges or problems to solve. The series uses project-based and problem-based learning, presenting students with the problem or challenge during the first lesson, and then teaching them science, social studies, English language arts, mathematics, and other content, as they apply what they learn to the challenge or problem at hand.

Authentic assessment and differentiation are embedded throughout the modules. Each *STEM Road Map Curriculum Series* module has a lead discipline, which may be science, social studies, English language arts, or mathematics. All disciplines are integrated into each module, along with ties to engineering. Another key component is the use of STEM Research Notebooks to allow students to track their own learning progress. The modules are designed with a scaffolded approach, with increasingly complex concepts and skills introduced as students progress through grade levels.

The developers of this work view the curriculum as a resource that is intended to be used either as a whole or in part to meet the needs of districts, schools, and teachers who are implementing an integrated STEM approach. A variety of implementation formats are possible, from using one stand-alone module at a given grade level to using all five modules to provide 25 weeks of instruction. Also, within each grade band (K–2, 3–5, 6–8, 9–12), the modules can be sequenced in various ways to suit specific needs.

STANDARDS-BASED APPROACH

The *STEM Road Map Curriculum Series* is anchored in the *Next Generation Science Standards* (*NGSS*), the *Common Core State Standards for Mathematics* (*CCSS Mathematics*), the *Common Core State Standards for English Language Arts* (*CCSS ELA*), and the Framework for 21st Century Learning. Each module includes a detailed curriculum map that incorporates the associated standards from the particular area correlated to lesson plans. The STEM Road Map has very clear and strong connections to these academic standards, and each of the grade-level topics was derived from the mapping of the standards to ensure alignment among topics, challenges or problems, and the required academic standards for students. Therefore, the curriculum series takes a standards-based approach and is designed to provide authentic contexts for application of required knowledge and skills.

THEMES IN THE *STEM ROAD MAP CURRICULUM SERIES*

The K–12 STEM Road Map is organized around five real-world STEM themes that were generated through an examination of the big ideas and challenges for society included in STEM standards and those that are persistent dilemmas for current and future generations:

- Cause and Effect
- Innovation and Progress
- The Represented World
- Sustainable Systems
- Optimizing the Human Experience

These themes are designed as springboards for launching students into an exploration of real-world learning situated within big ideas. Most important, the five STEM Road Map themes serve as a framework for scaffolding STEM learning across the K–12 continuum.

The themes are distributed across the STEM disciplines so that they represent the big ideas in science (Cause and Effect; Sustainable Systems), technology (Innovation and Progress; Optimizing the Human Experience), engineering (Innovation and Progress; Sustainable Systems; Optimizing the Human Experience), and mathematics (The Represented World), as well as concepts and challenges in social studies and 21st century skills that are also excellent contexts for learning in English language arts. The process of developing themes began with the clustering of the *NGSS* performance expectations and the National Academy of Engineering's grand challenges for engineering, which led to the development of the challenge in each module and connections of the module activities to the *CCSS Mathematics* and *CCSS ELA* standards. We performed these mapping processes with large teams of experts and found that these five themes provided breadth, depth, and coherence to frame a high-quality STEM learning experience from kindergarten through 12th grade.

Cause and Effect

The concept of cause and effect is a powerful and pervasive notion in the STEM fields. It is the foundation of understanding how and why things happen as they do. Humans spend considerable effort and resources trying to understand the causes and effects of natural and designed phenomena to gain better control over events and the environment and to be prepared to react appropriately. Equipped with the knowledge of a specific cause-and-effect relationship, we can lead better lives or contribute to the community by altering the cause, leading to a different effect. For example, if a person recognizes that irresponsible energy consumption leads to global climate change, that person can act to remedy his or her contribution to the situation. Although cause and effect is a core idea in the STEM fields, it can actually be difficult to determine. Students should be capable of understanding not only when evidence points to cause and effect but also when evidence points to relationships but not direct causality. The major goal of education is to foster students to be empowered, analytic thinkers, capable of thinking through complex processes to make important decisions. Understanding causality, as well as when it cannot be determined, will help students become better consumers, global citizens, and community members.

Innovation and Progress

One of the most important factors in determining whether humans will have a positive future is innovation. Innovation is the driving force behind progress, which helps create possibilities that did not exist before. Innovation and progress are creative entities, but in the STEM fields, they are anchored by evidence and logic, and they use established concepts to move the STEM fields forward. In creating something new, students must consider what is already known in the STEM fields and apply this knowledge appropriately. When we innovate, we create value that was not there previously and create new conditions and possibilities for even more innovations. Students should consider how their innovations might affect progress and use their STEM thinking to change current human burdens to benefits. For example, if we develop more efficient cars that use by-products from another manufacturing industry, such as food processing, then we have used waste productively and reduced the need for the waste to be hauled away, an indirect benefit of the innovation.

The Represented World

When we communicate about the world we live in, how the world works, and how we can meet the needs of humans, sometimes we can use the actual phenomena to explain a concept. Sometimes, however, the concept is too big, too slow, too small, too fast, or too complex for us to explain using the actual phenomena, and we must use a representation or a model to help communicate the important features. We need representations and models such as graphs, tables, mathematical expressions, and diagrams because it makes our thinking visible. For example, when examining geologic time, we cannot actually observe the passage of such large chunks of time, so we create a timeline or a model that uses a proportional scale to visually illustrate how much time has passed for different eras. Another example may be something too complex for students at a particular grade level, such as explaining the *p* subshell orbitals of electrons to fifth graders. Instead, we use the Bohr model, which more closely represents the orbiting of planets and is accessible to fifth graders.

When we create models, they are helpful because they point out the most important features of a phenomenon. We also create representations of the world with mathematical functions, which help us change parameters to suit the situation. Creating representations of a phenomenon engages students because they are able to identify the important features of that phenomenon and communicate them directly. But because models are estimates of a phenomenon, they leave out some of the details, so it is important for students to evaluate their usefulness as well as their shortcomings.

Sustainable Systems

From an engineering perspective, the term *system* refers to the use of "concepts of component need, component interaction, systems interaction, and feedback. The interaction of subcomponents to produce a functional system is a common lens used by all engineering disciplines for understanding, analysis, and design." (Koehler, Bloom, and Binns 2013, p. 8). Systems can be either open (e.g., an ecosystem) or closed (e.g., a car battery). Ideally, a system should be sustainable, able to maintain equilibrium without much energy from outside the structure. Looking at a garden, we see flowers blooming, weeds sprouting, insects buzzing, and various forms of life living within its boundaries. This is an example of an ecosystem, a collection of living organisms that survive together, functioning as a system. The interaction of the organisms within the system and the influences of the environment (e.g., water, sunlight) can maintain the system for a period of time, thus demonstrating its ability to endure. Sustainability is a desirable feature of a system because it allows for existence of the entity in the long term.

In the STEM Road Map project, we identified different standards that we consider to be oriented toward systems that students should know and understand in the K–12 setting. These include ecosystems, the rock cycle, Earth processes (such as erosion, tectonics, ocean currents, weather phenomena), Earth-Sun-Moon cycles, heat transfer, and the interaction among the geosphere, biosphere, hydrosphere, and atmosphere. Students and teachers should understand that we live in a world of systems that are not independent of each other, but rather are intrinsically linked such that a disruption in one part of a system will have reverberating effects on other parts of the system.

Optimizing the Human Experience

Science, technology, engineering, and mathematics as disciplines have the capacity to continuously improve the ways humans live, interact, and find meaning in the world, thus working to optimize the human experience. This idea has two components: being more suited to our environment and being more fully human. For example, the progression of STEM ideas can help humans create solutions to complex problems, such as improving ways to access water sources, designing energy sources with minimal impact on our environment, developing new ways of communication and expression, and building efficient shelters. STEM ideas can also provide access to the secrets and wonders of nature. Learning in STEM requires students to think logically and systematically, which is a way of knowing the world that is markedly different from knowing the world as an artist. When students can employ various ways of knowing and understand when it is appropriate to use a different way of knowing or integrate ways of knowing, they are fully experiencing the best of what it is to be human. The problem-based learning scenarios provided in the STEM Road Map help students develop ways of thinking like STEM professionals as they ask questions and design solutions. They learn to optimize the human experience by innovating improvements in the designed world in which they live.

THE NEED FOR AN INTEGRATED STEM APPROACH

At a basic level, STEM stands for science, technology, engineering, and mathematics. Over the past decade, however, STEM has evolved to have a much broader scope and implications. Now, educators and policy makers refer to STEM as not only a concentrated area for investing in the future of the United States and other nations but also as a domain and mechanism for educational reform.

The good intentions of the recent decade-plus of focus on accountability and increased testing has resulted in significant decreases not only in instructional time for teaching science and social studies but also in the flexibility of teachers to promote authentic, problem solving–focused classroom environments. The shift has had a detrimental impact on student acquisition of vitally important skills, which many refer to as 21st century skills, and often the ability of students to "think." Further, schooling has become increasingly siloed into compartments of mathematics, science, English language arts, and social studies, lacking any of the connections that are overwhelmingly present in

the real world around children. Students have experienced school as content provided in boxes that must be memorized, devoid of any real-world context, and often have little understanding of why they are learning these things.

STEM-focused projects, curriculum, activities, and schools have emerged as a means to address these challenges. However, most of these efforts have continued to focus on the individual STEM disciplines (predominantly science and engineering) through more STEM classes and after-school programs in a "STEM enhanced" approach (Breiner et al. 2012). But in traditional and STEM enhanced approaches, there is little to no focus on other disciplines that are integral to the context of STEM in the real world. Integrated STEM education, on the other hand, infuses the learning of important STEM content and concepts with a much-needed emphasis on 21st century skills and a problem- and project-based pedagogy that more closely mirrors the real-world setting for society's challenges. It incorporates social studies, English language arts, and the arts as pivotal and necessary (Johnson 2013; Rennie, Venville, and Wallace 2012; Roehrig et al. 2012).

FRAMEWORK FOR STEM INTEGRATION IN THE CLASSROOM

The *STEM Road Map Curriculum Series* is grounded in the Framework for STEM Integration in the Classroom as conceptualized by Moore, Guzey, and Brown (2014) and Moore et al. (2014). The framework has six elements, described in the context of how they are used in the *STEM Road Map Curriculum Series* as follows:

1. The STEM Road Map contexts are meaningful to students and provide motivation to engage with the content. Together, these allow students to have different ways to enter into the challenge.

2. The STEM Road Map modules include engineering design that allows students to design technologies (i.e., products that are part of the designed world) for a compelling purpose.

3. The STEM Road Map modules provide students with the opportunities to learn from failure and redesign based on the lessons learned.

4. The STEM Road Map modules include standards-based disciplinary content as the learning objectives.

5. The STEM Road Map modules include student-centered pedagogies that allow students to grapple with the content, tie their ideas to the context, and learn to think for themselves as they deepen their conceptual knowledge.

6. The STEM Road Map modules emphasize 21st century skills and, in particular, highlight communication and teamwork.

All of the STEM Road Map modules incorporate these six elements; however, the level of emphasis on each of these elements varies based on the challenge or problem in each module.

THE NEED FOR THE *STEM ROAD MAP CURRICULUM SERIES*

As focus is increasing on integrated STEM, and additional schools and programs decide to move their curriculum and instruction in this direction, there is a need for high-quality, research-based curriculum designed with integrated STEM at the core. Several good resources are available to help teachers infuse engineering or more STEM enhanced approaches, but no curriculum exists that spans K–12 with an integrated STEM focus. The next chapter provides detailed information about the specific pedagogy, instructional strategies, and learning theory on which the *STEM Road Map Curriculum Series* is grounded.

REFERENCES

Breiner, J., M. Harkness, C. C. Johnson, and C. Koehler. 2012. What is STEM? A discussion about conceptions of STEM in education and partnerships. *School Science and Mathematics* 112 (1): 3–11.

Johnson, C. C. 2013. Conceptualizing integrated STEM education: Editorial. *School Science and Mathematics* 113 (8): 367–368.

Koehler, C. M., M. A. Bloom, and I. C. Binns. 2013. Lights, camera, action: Developing a methodology to document mainstream films' portrayal of nature of science and scientific inquiry. *Electronic Journal of Science Education* 17 (2).

Moore, T. J., S. S. Guzey, and A. Brown. 2014. Greenhouse design to increase habitable land: An engineering unit. *Science Scope* 51–57.

Moore, T. J., M. S. Stohlmann, H.-H. Wang, K. M. Tank, A. W. Glancy, and G. H. Roehrig. 2014. Implementation and integration of engineering in K–12 STEM education. In *Engineering in pre-college settings: Synthesizing research, policy, and practices,* ed. S. Purzer, J. Strobel, and M. Cardella, 35–60. West Lafayette, IN: Purdue Press.

Rennie, L., G. Venville, and J. Wallace. 2012. *Integrating science, technology, engineering, and mathematics: Issues, reflections, and ways forward.* New York: Routledge.

Roehrig, G. H., T. J. Moore, H. H. Wang, and M. S. Park. 2012. Is adding the *E* enough? Investigating the impact of K–12 engineering standards on the implementation of STEM integration. *School Science and Mathematics* 112 (1): 31–44.

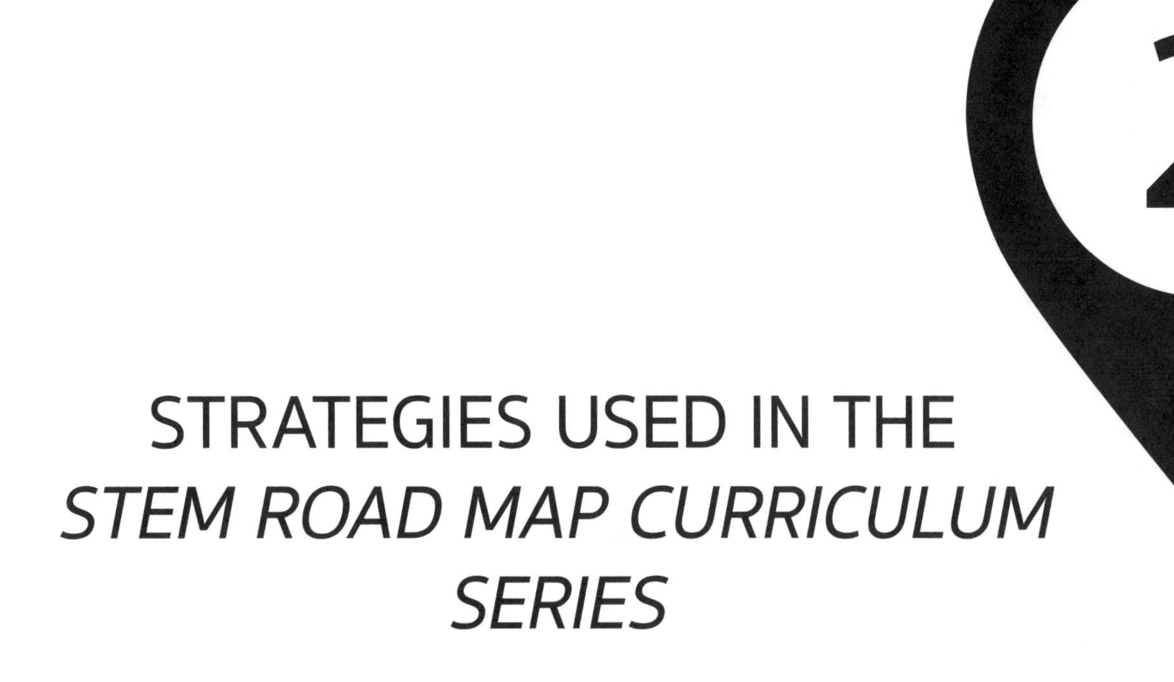

STRATEGIES USED IN THE *STEM ROAD MAP CURRICULUM SERIES*

Erin Peters-Burton, Carla C. Johnson, Toni A. Sondergeld, and Tamara J. Moore

The *STEM Road Map Curriculum Series* uses what has been identified through research as best-practice pedagogy, including embedded formative assessment strategies throughout each module. This chapter briefly describes the key strategies that are employed in the series.

PROJECT- AND PROBLEM-BASED LEARNING

Each module in the *STEM Road Map Curriculum Series* uses either project-based learning or problem-based learning to drive the instruction. Project-based learning begins with a driving question to guide student teams in addressing a contextualized local or community problem or issue. The outcome of project-based instruction is a product that is conceptualized, designed, and tested through a series of scaffolded learning experiences (Blumenfeld et al. 1991; Krajcik and Blumenfeld 2006). Problem-based learning is often grounded in a fictitious scenario, challenge, or problem (Barell 2006; Lambros 2004). On the first day of instruction within the unit, student teams are provided with the context of the problem. Teams work through a series of activities and use open-ended research to develop their potential solution to the problem or challenge, which need not be a tangible product (Johnson 2003).

ENGINEERING DESIGN PROCESS

The *STEM Road Map Curriculum Series* uses engineering design as a way to facilitate integrated STEM within the modules. The engineering design process (EDP) is depicted in Figure 2.1 (p. 10). It highlights two major aspects of engineering design—problem scoping and solution generation—and six specific components of working toward a design: define the problem, learn about the problem, plan a solution, try the solution, test the solution, decide whether the solution is good enough. It also shows that communication

Figure 2.1. Engineering Design Process

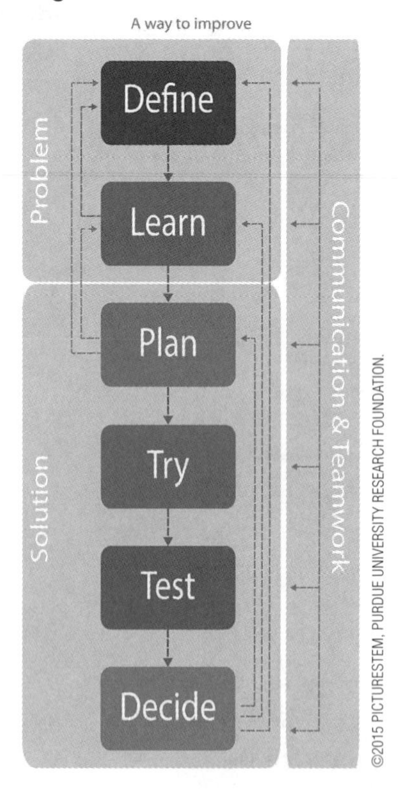

and teamwork are involved throughout the entire process. As the arrows in the figure indicate, the order in which the components of engineering design are addressed depends on what becomes needed as designers progress through the EDP. Designers must communicate and work in teams throughout the process. The EDP is iterative, meaning that components of the process can be repeated as needed until the design is good enough to present to the client as a potential solution to the problem.

Problem scoping is the process of gathering and analyzing information to deeply understand the engineering design problem. It includes defining the problem and learning about the problem. Defining the problem includes identifying the problem, the client, and the end user of the design. The client is the person (or people) who hired the designers to do the work, and the end user is the person (or people) who will use the final design. The designers must also identify the criteria and the constraints of the problem. The criteria are the things the client wants from the solution, and the constraints are the things that limit the possible solutions. The designers must spend significant time learning about the problem, which can include activities such as the following:

- Reading informational texts and researching about relevant concepts or contexts
- Identifying and learning about needed mathematical and scientific skills, knowledge, and tools
- Learning about things done previously to solve similar problems
- Experimenting with possible materials that could be used in the design

Problem scoping also allows designers to consider how to measure the success of the design in addressing specific criteria and staying within the constraints over multiple iterations of solution generation.

Solution generation includes planning a solution, trying the solution, testing the solution, and deciding whether the solution is good enough. Planning the solution includes generating many design ideas that both address the criteria and meet the constraints. Here the designers must consider what was learned about the problem during problem scoping. Design plans include clear communication of design ideas through media such as notebooks, blueprints, schematics, or storyboards. They also include details about the design, such as measurements, materials, colors, costs of materials, instructions for how things fit together, and sets of directions. Making the decision about which design idea to move forward involves considering the trade-offs of each design idea.

Once a clear design plan is in place, the designers must try the solution. Trying the solution includes developing a prototype (a testable model) based on the plan generated. The prototype might be something physical or a process to accomplish a goal. This component of design requires that the designers consider the risk involved in implementing the design. The prototype developed must be tested. Testing the solution includes conducting fair tests that verify whether the plan is a solution that is good enough to meet the client and end user needs and wants. Data need to be collected about the results of the tests of the prototype, and these data should be used to make evidence-based decisions regarding the design choices made in the plan. Here, the designers must again consider the criteria and constraints for the problem.

Using the data gathered from the testing, the designers must decide whether the solution is good enough to meet the client and end user needs and wants by assessment based on the criteria and constraints. Here, the designers must justify or reject design decisions based on the background research gathered while learning about the problem and on the evidence gathered during the testing of the solution. The designers must now decide whether to present the current solution to the client as a possibility or to do more iterations of design on the solution. If they decide that improvements need to be made to the solution, the designers must decide if there is more that needs to be understood about the problem, client, or end user; if another design idea should be tried; or if more planning needs to be conducted on the same design. One way or another, more work needs to be done.

Throughout the process of designing a solution to meet a client's needs and wants, designers work in teams and must communicate to each other, the client, and likely the end user. Teamwork is important in engineering design because multiple perspectives and differing skills and knowledge are valuable when working to solve problems. Communication is key to the success of the designed solution. Designers must communicate their ideas clearly using many different representations, such as text in an engineering notebook, diagrams, flowcharts, technical briefs, or memos to the client.

LEARNING CYCLE

The same format for the learning cycle is used in all grade levels throughout the STEM Road Map, so that students engage in a variety of activities to learn about phenomena in the modules thoroughly and have consistent experiences in the problem- and project-based learning modules. Expectations for learning by younger students are not as high as for older students, but the format of the progression of learning is the same. Students who have learned with curriculum from the STEM Road Map in early grades know

what to expect in later grades. The learning cycle consists of five parts—Introductory Activity/Engagement, Activity/Exploration, Explanation, Elaboration/Application of Knowledge, and Evaluation/Assessment—and is based on the empirically tested 5E model from BSCS (Bybee et al. 2006).

In the Introductory Activity/Engagement phase, teachers introduce the module challenge and use a unique approach designed to pique students' curiosity. This phase gets students to start thinking about what they already know about the topic and begin wondering about key ideas. The Introductory Activity/Engagement phase positions students to be confident about what they are about to learn, because they have prior knowledge, and clues them into what they don't yet know.

In the Activity/Exploration phase, the teacher sets up activities in which students experience a deeper look at the topics that were introduced earlier. Students engage in the activities and generate new questions or consider possibilities using preliminary investigations. Students work independently, in small groups, and in whole-group settings to conduct investigations, resulting in common experiences about the topic and skills involved in the real-world activities. Teachers can assess students' development of concepts and skills based on the common experiences during this phase.

During the Explanation phase, teachers direct students' attention to concepts they need to understand and skills they need to possess to accomplish the challenge. Students participate in activities to demonstrate their knowledge and skills to this point, and teachers can pinpoint gaps in student knowledge during this phase.

In the Elaboration/Application of Knowledge phase, teachers present students with activities that engage in higher-order thinking to create depth and breadth of student knowledge, while connecting ideas across topics within and across STEM. Students apply what they have learned thus far in the module to a new context or elaborate on what they have learned about the topic to a deeper level of detail.

In the last phase, Evaluation/Assessment, teachers give students summative feedback on their knowledge and skills as demonstrated through the challenge. This is not the only point of assessment (as discussed in the section on Embedded Formative Assessments), but it is an assessment of the culmination of the knowledge and skills for the module. Students demonstrate their cognitive growth at this point and reflect on how far they have come since the beginning of the module. The challenges are designed to be multidimensional in the ways students must collaborate and communicate their new knowledge.

STEM RESEARCH NOTEBOOK

One of the main components of the *STEM Road Map Curriculum Series* is the STEM Research Notebook, a place for students to capture their ideas, questions, observations, reflections, evidence of progress, and other items associated with their daily work. At the beginning of each module, the teacher walks students through the setup of the STEM

Research Notebook, which could be a three-ring binder, composition book, or spiral notebook. You may wish to have students create divided sections so that they can easily access work from various disciplines during the module. Electronic notebooks kept on student devices are also acceptable and encouraged. Students will develop their own table of contents and create chapters in the notebook for each module.

Each lesson in the *STEM Road Map Curriculum Series* includes one or more prompts that are designed for inclusion in the STEM Research Notebook and appear as questions or statements that the teacher assigns to students. These prompts require students to apply what they have learned across the lesson to solve the big problem or challenge for that module. Each lesson is designed to meaningfully refer students to the larger problem or challenge they have been assigned to solve with their teams. The STEM Research Notebook is designed to be a key formative assessment tool, as students' daily entries provide evidence of what they are learning. The notebook can be used as a mechanism for dialogue between the teacher and students, as well as for peer and self-evaluation.

The use of the STEM Research Notebook is designed to scaffold student notebooking skills across the grade bands in the *STEM Road Map Curriculum Series*. In the early grades, children learn how to organize their daily work in the notebook as a way to collect their products for future reference. In elementary school, students structure their notebooks to integrate background research along with their daily work and lesson prompts. In the upper grades (middle and high school), students expand their use of research and data gathering through team discussions to more closely mirror the work of STEM experts in the real world.

THE ROLE OF ASSESSMENT IN THE *STEM ROAD MAP CURRICULUM SERIES*

Starting in the middle years and continuing into secondary education, the word *assessment* typically brings grades to mind. These grades may take the form of a letter or a percentage, but they typically are used as a representation of a student's content mastery. If well thought out and implemented, however, classroom assessment can offer teachers, parents, and students valuable information about student learning and misconceptions that does not necessarily come in the form of a grade (Popham 2013).

The *STEM Road Map Curriculum Series* provides a set of assessments for each module. Teachers are encouraged to use assessment information for more than just assigning grades to students. Instead, assessments of activities requiring students to actively engage in their learning, such as student journaling in STEM Research Notebooks, collaborative presentations, and constructing graphic organizers, should be used to move student learning forward. Whereas other curriculum with assessments may include objective-type (multiple-choice or matching) tests, quizzes, or worksheets, we have intentionally avoided these forms of assessments to better align assessment strategies with teacher instruction and

student learning techniques. Since the focus of this book is on project- or problem-based STEM curriculum and instruction that focuses on higher-level thinking skills, appropriate and authentic performance assessments were developed to elicit the most reliable and valid indication of growth in student abilities (Brookhart and Nitko 2008).

Comprehensive Assessment System

Assessment throughout all STEM Road Map curriculum modules acts as a comprehensive system in which formative and summative assessments work together to provide teachers with high-quality information on student learning. Formative assessment occurs when the teacher finds out formally or informally what a student knows about a smaller, defined concept or skill and provides timely feedback to the student about his or her level of proficiency. Summative assessments occur when students have performed all activities in the module and are given a cumulative performance evaluation in which they demonstrate their growth in learning.

A comprehensive assessment system can be thought of as akin to a sporting event. Formative assessments are the practices: It is important to accomplish them consistently, they provide feedback to help students improve their learning, and making mistakes can be worthwhile if students are given an opportunity to learn from them. Summative assessments are the competitions: Students need to be prepared to perform at the best of their ability. Without multiple opportunities to practice skills along the way through formative assessments, students will not have the best chance of demonstrating growth in abilities through summative assessments (Black and Wiliam 1998).

Embedded Formative Assessments

Formative assessments in this module serve two main purposes: to provide feedback to students about their learning and to provide important information for the teacher to inform immediate instructional needs. Providing feedback to students is particularly important when conducting problem- or project-based learning because students take on much of the responsibility for learning, and teachers must facilitate student learning in an informed way. For example, if students are required to conduct research for the Activity/Exploration phase but are not familiar with what constitutes a reliable resource, they may develop misconceptions based on poor information. When a teacher monitors this learning through formative assessments and provides specific feedback related to the instructional goals, students are less likely to develop incomplete or incorrect conceptions in their independent investigations. By using formative assessment to detect problems in student learning and then acting on this information, teachers help move student learning forward through these teachable moments.

Formative assessments come in a variety of formats. They can be informal, such as asking students probing questions related to student knowledge or tasks or simply observing students engaged in an activity to gather information about student skills. Formative assessments can also be formal, such as a written quiz or a laboratory practical. Regardless of the type, three key steps must be completed when using formative assessments (Sondergeld, Bell, and Leusner 2010). First, the assessment is delivered to students so that teachers can collect data. Next, teachers analyze the data (student responses) to determine student strengths and areas that need additional support. Finally, teachers use the results from information collected to modify lessons and create learning environments that reinforce weak points in student learning. If student learning information is not used to modify instruction, the assessment cannot be considered formative in nature.

Formative assessments can be about content, science process skills, or even learning skills. When a formative assessment focuses on content, it assesses student knowledge about the disciplinary core ideas from the *Next Generation Science Standards* (*NGSS*) or content objectives from *Common Core State Standards for Mathematics* (*CCSS Mathematics*) or *Common Core State Standards for English Language Arts* (*CCSS ELA*). Content-focused formative assessments ask students questions about declarative knowledge regarding the concepts they have been learning. Process skills formative assessments examine the extent to which a student can perform science and engineering practices from the *NGSS* or process objectives from *CCSS Mathematics* or *CCSS ELA*, such as constructing an argument. Learning skills can also be assessed formatively by asking students to reflect on the ways they learn best during a module and identify ways they could have learned more.

Assessment Maps

Assessment maps or blueprints can be used to ensure alignment between classroom instruction and assessment. If what students are learning in the classroom is not the same as the content on which they are assessed, the resultant judgment made on student learning will be invalid (Brookhart and Nitko 2008). Therefore, the issue of instruction and assessment alignment is critical. The assessment map for this book (found in Chapter 3) indicates by lesson whether the assessment should be completed as a group or on an individual basis, identifies the assessment as formative or summative in nature, and aligns the assessment with its corresponding learning objectives.

Note that the module includes far more formative assessments than summative assessments. This is done intentionally to provide students with multiple opportunities to practice their learning of new skills before completing a summative assessment. Note also that formative assessments are used to collect information on only one or two learning objectives at a time so that potential relearning or instructional modifications can focus on smaller and more manageable chunks of information. Conversely, summative assessments in the module cover many more learning objectives, as they are traditionally used as final markers of student learning. This is not to say that information collected from summative assessments cannot or should not be used formatively. If teachers find that gaps in student

learning persist after a summative assessment is completed, it is important to revisit these existing misconceptions or areas of weakness before moving on (Black et al. 2003).

SELF-REGULATED LEARNING THEORY IN THE STEM ROAD MAP MODULES

Many learning theories are compatible with the STEM Road Map modules, such as constructivism, situated cognition, and meaningful learning. However, we feel that the self-regulated learning theory (SRL) aligns most appropriately (Zimmerman 2000). SRL requires students to understand that thinking needs to be motivated and managed (Ritchhart, Church, and Morrison 2011). The STEM Road Map modules are student centered and are designed to provide students with choices, concrete hands-on experiences, and opportunities to see and make connections, especially across subjects (Eliason and Jenkins 2012; NAEYC 2016). Additionally, SRL is compatible with the modules because it fosters a learning environment that supports students' motivation, enables students to become aware of their own learning strategies, and requires reflection on learning while experiencing the module (Peters and Kitsantas 2010).

The theory behind SRL (see Figure 2.2) explains the different processes that students engage in before, during, and after a learning task. Because SRL is a cyclical learning process, the accomplishment of one cycle develops strategies for the next learning cycle. This cyclic way of learning aligns with the various sections in the STEM Road Map lesson plans on Introductory Activity/Engagement, Activity/Exploration, Explanation, Elaboration/Application of Knowledge, and Evaluation/Assessment. Since the students engaged in a module take on much of the responsibility for learning, this theory also provides guidance for teachers to keep students on the right track.

Figure 2.2. SRL Theory

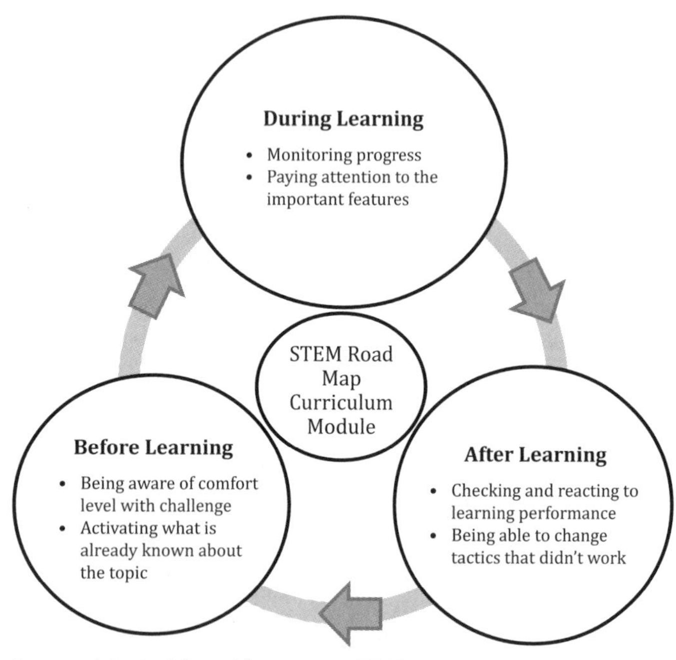

Source: Adapted from Zimmerman 2000.

The remainder of this section explains how SRL theory is embedded within the five sections of each module and points out ways to support students in becoming independent learners of STEM while productively functioning in collaborative teams.

Before Learning: Setting the Stage

Before attempting a learning task such as the STEM Road Map modules, teachers should develop an understanding of their students' level of comfort with the process of accomplishing the learning and determine what they already know about the topic. When students are comfortable with attempting a learning task, they tend to take more risks in learning and as a result achieve deeper learning (Bandura 1986).

The STEM Road Map curriculum modules are designed to foster excitement from the very beginning. Each module has an Introductory Activity/Engagement section that introduces the overall topic from a unique and exciting perspective, engaging the students to learn more so that they can accomplish the challenge. The Introductory Activity also has a design component that helps teachers assess what students already know about the topic of the module. In addition to the deliberate designs in the lesson plans to support SRL, teachers can support a high level of student comfort with the learning challenge by finding out if students have ever accomplished the same kind of task and, if so, asking them to share what worked well for them.

During Learning: Staying the Course

Some students fear inquiry learning because they aren't sure what to do to be successful (Peters 2010). However, the STEM Road Map curriculum modules are embedded with tools to help students pay attention to knowledge and skills that are important for the learning task and to check student understanding along the way. One of the most important processes for learning is the ability for learners to monitor their own progress while performing a learning task (Peters 2012). The modules allow students to monitor their progress with tools such as the STEM Research Notebooks, in which they record what they know and can check whether they have acquired a complete set of knowledge and skills. The STEM Road Map modules support inquiry strategies that include previewing, questioning, predicting, clarifying, observing, discussing, and journaling (Morrison and Milner 2014). Through the use of technology throughout the modules, inquiry is supported by providing students access to resources and data while enabling them to process information, report the findings, collaborate, and develop 21st century skills.

It is important for teachers to encourage students to have an open mind about alternative solutions and procedures (Milner and Sondergeld 2015) when working through the STEM Road Map curriculum modules. Novice learners can have difficulty knowing what to pay attention to and tend to treat each possible avenue for information as equal (Benner 1984). Teachers are the mentors in a classroom and can point out ways for students to approach learning during the Activity/Exploration, Explanation, and

Elaboration/Application of Knowledge portions of the lesson plans to ensure that students pay attention to the important concepts and skills throughout the module. For example, if a student is to demonstrate conceptual awareness of motion when working on roller coaster research, but the student has misconceptions about motion, the teacher can step in and redirect student learning.

After Learning: Knowing What Works

The classroom is a busy place, and it may often seem that there is no time for self-reflection on learning. Although skipping this reflective process may save time in the short term, it reduces the ability to take into account things that worked well and things that didn't so that teaching the module may be improved next time. In the long run, SRL skills are critical for students to become independent learners who can adapt to new situations. By investing the time it takes to teach students SRL skills, teachers can save time later, because students will be able to apply methods and approaches for learning that they have found effective to new situations. In the Evaluation/Assessment portion of the STEM Road Map curriculum modules, as well as in the formative assessments throughout the modules, two processes in the after-learning phase are supported: evaluating one's own performance and accounting for ways to adapt tactics that didn't work well. Students have many opportunities to self-assess in formative assessments, both in groups and individually, using the rubrics provided in the modules.

The designs of the *NGSS* and *CCSS* allow for students to learn in diverse ways, and the STEM Road Map curriculum modules emphasize that students can use a variety of tactics to complete the learning process. For example, students can use STEM Research Notebooks to record what they have learned during the various research activities. Notebook entries might include putting objectives in students' own words, compiling their prior learning on the topic, documenting new learning, providing proof of what they learned, and reflecting on what they felt successful doing and what they felt they still needed to work on. Perhaps students didn't realize that they were supposed to connect what they already knew with what they learned. They could record this and would be prepared in the next learning task to begin connecting prior learning with new learning.

SAFETY IN STEM

Student safety is a primary consideration in all subjects but is an area of particular concern in science, where students may interact with unfamiliar tools and materials that may pose additional safety risks. It is important to implement safety practices within the context of STEM investigations, whether in a classroom laboratory or in the field. When you keep safety in mind as a teacher, you avoid many potential issues with the lesson while also protecting your students.

STEM safety practices encompass things considered in the typical science classroom. Ensure that students are familiar with basic safety considerations, such as wearing protective equipment (e.g., safety glasses or goggles and latex-free gloves) and taking care with sharp objects, and know emergency exit procedures. Teachers should learn beforehand the locations of the safety eyewash, fume hood, fire extinguishers, and emergency shut-off switch in the classroom and how to use them. Also be aware of any school or district safety policies that are in place and apply those that align with the work being conducted in the lesson. It is important to review all safety procedures annually.

STEM investigations should always be supervised. Each lesson in the modules includes teacher guidelines for applicable safety procedures that should be followed. Before each investigation, teachers should go over these safety procedures with the student teams. Some STEM focus areas such as engineering require that students can demonstrate how to properly use equipment in the maker space before the teacher allows them to proceed with the lesson.

Information about classroom science safety, including a safety checklist for science classrooms, general lab safety recommendations, and links to other science safety resources, is available at the Council of State Science Supervisors (CSSS) website at *www.csss-science.org/safety.shtml.* The National Science Teachers Association (NSTA) provides a list of science rules and regulations, including standard operating procedures for lab safety, and a safety acknowledgement form for students and parents or guardians to sign. You can access these resources at *http://static.nsta.org/pdfs/SafetyInTheScienceClassroom.pdf.* In addition, NSTA's Safety in the Science Classroom web page (*www.nsta.org/safety*) has numerous links to safety resources, including papers written by the NSTA Safety Advisory Board.

Disclaimer: The safety precautions for each activity are based on use of the recommended materials and instructions, legal safety standards, and better professional practices. Using alternative materials or procedures for these activities may jeopardize the level of safety and therefore is at the user's own risk.

REFERENCES

Bandura, A. 1986. *Social foundations of thought and action: A social cognitive theory.* Englewood Cliffs, NJ: Prentice-Hall.

Barell, J. 2006. *Problem-based learning: An inquiry approach.* Thousand Oaks, CA: Corwin Press.

Benner, P. 1984. *From novice to expert: Excellence and power in clinical nursing practice.* Menlo Park, CA: Addison-Wesley Publishing Company.

Black, P., C. Harrison, C. Lee, B. Marshall, and D. Wiliam. 2003. *Assessment for learning: Putting it into practice.* Berkshire, UK: Open University Press.

Black, P., and D. Wiliam. 1998. Inside the black box: Raising standards through classroom assessment. *Phi Delta Kappan* 80 (2): 139–148.

Blumenfeld, P., E. Soloway, R. Marx, J. Krajcik, M. Guzdial, and A. Palincsar. 1991. Motivating project-based learning: Sustaining the doing, supporting learning. *Educational Psychologist* 26 (3): 369–398.

Brookhart, S. M., and A. J. Nitko. 2008. *Assessment and grading in classrooms.* Upper Saddle River, NJ: Pearson.

Bybee, R., J. Taylor, A. Gardner, P. Van Scotter, J. Carlson, A. Westbrook, And N. Landes. 2006. *The BSCS 5E instructional model: Origins and effectiveness. http://science.education.nih.gov/houseofreps.nsf/b82d55fa138783c2852572c9004f5566/$FILE/Appendix?D.pdf.*

Eliason, C. F., and L. T. Jenkins. 2012. *A practical guide to early childhood curriculum.* 9th ed. New York: Merrill.

Johnson, C. 2003. Bioterrorism is real-world science: Inquiry-based simulation mirrors real life. *Science Scope* 27 (3): 19–23.

Krajcik, J., and P. Blumenfeld. 2006. Project-based learning. In *The Cambridge handbook of the learning sciences,* ed. R. Keith Sawyer, 317–334. New York: Cambridge University Press.

Lambros, A. 2004. *Problem-based learning in middle and high school classrooms: A teacher's guide to implementation.* Thousand Oaks, CA: Corwin Press.

Milner, A. R., and T. Sondergeld. 2015. Gifted urban middle school students: The inquiry continuum and the nature of science. *National Journal of Urban Education and Practice* 8 (3): 442–461.

Morrison, V., and A. R. Milner. 2014. Literacy in support of science: A closer look at cross-curricular instructional practice. *Michigan Reading Journal* 46 (2): 42–56.

National Association for the Education of Young Children (NAEYC). 2016. Developmentally appropriate practice position statements. *www.naeyc.org/positionstatements/dap.*

Peters, E. E. 2010. Shifting to a student-centered science classroom: An exploration of teacher and student changes in perceptions and practices. *Journal of Science Teacher Education* 21 (3): 329–349.

Peters, E. E. 2012. Developing content knowledge in students through explicit teaching of the nature of science: Influences of goal setting and self-monitoring. *Science and Education* 21 (6): 881–898.

Peters, E. E., and A. Kitsantas. 2010. The effect of nature of science metacognitive prompts on science students' content and nature of science knowledge, metacognition, and self-regulatory efficacy. *School Science and Mathematics* 110: 382–396.

Popham, W. J. 2013. *Classroom assessment: What teachers need to know.* 7th ed. Upper Saddle River, NJ: Pearson.

Ritchhart, R., M. Church, and K. Morrison. 2011. *Making thinking visible: How to promote engagement, understanding, and independence for all learners.* San Francisco, CA: Jossey-Bass.

Sondergeld, T. A., C. A. Bell, and D. M. Leusner. 2010. Understanding how teachers engage in formative assessment. *Teaching and Learning* 24 (2): 72–86.

Zimmerman, B. J. 2000. Attaining self-regulation: A social-cognitive perspective. In *Handbook of self-regulation,* ed. M. Boekaerts, P. Pintrich, and M. Zeidner, 13–39. San Diego: Academic Press.

IMPROVING BRIDGE DESIGN

STEM ROAD MAP MODULE

IMPROVING BRIDGE DESIGN MODULE OVERVIEW

John Weaver, Toni A. Ivey, Juliana Utley, Adrienne Redmond-Sanogo, Sue Christian Parsons, Janet B. Walton, Carla C. Johnson, and Erin Peters-Burton

THEME: The Represented World

LEAD DISCIPLINE: Mathematics

MODULE SUMMARY

This module focuses on addressing the real problems of today's society through the lens of the past. The challenge for this module is led by mathematics and is focused on infrastructure decay, specifically the state of bridges in the United States. With recent bridge collapses (e.g., the Minnesota bridge in 2007), much debate has ensued about the maintenance of bridges, and designs that will prove to be more sustainable over time are now being examined. Student teams develop a decision model grounded in engineering, for the local department of transportation, on how to select bridge design aligned with appropriate span length, application, use information, and other important data. In science, students examine observable changes in rocks and fossils to interpret the past. In English language arts (ELA), students work to develop a written proposal that articulates key components of their decision model (Johnson et al., 2015, p. 116). In social studies, students learn about how infrastructure such as roads and bridges has helped move their geographic region forward. (*Note:* This module instructs teachers to show videos of collapsing bridges. Teachers should consider students' sensitivity to the videos before showing them.)

ESTABLISHED GOALS AND OBJECTIVES

At the conclusion of this module, students will be able to do the following:

- Use mathematical modeling to explore bridge design, structure, and function, as well as to develop a decision model to help a community make appropriate decisions that will have a positive impact on their local infrastructure. (Mathematics)

- Understand how Earth materials play an important role in all aspects of our modern lives, including the construction of roadways and bridges. (Science)
- Employ research, nonfiction writing, and multimodal composition skills to explore and communicate the significance of bridges in our cultural experiences and understandings. (ELA)
- Investigate how infrastructure such as roads and bridges affect individual and local culture. (Social Studies)
- Build mastery of relevant skills and themes of the Framework for 21st Century Learning.

CHALLENGE OR PROBLEM FOR STUDENTS TO SOLVE: BRIDGE DESIGN CHALLENGE

The teacher should explain the challenge to the students as follows: Because of the current state of bridges in the United States, we are going to spend the next few weeks researching, designing, testing, and constructing bridges. Our challenge is to help the local department of transportation make better choices that will have a positive impact on our nation's infrastructure. By making better decisions we can help ensure that future bridges are sustainable and appropriate for the community in which they are built.

As we discuss the variables involved in building a bridge, you will be working in groups to develop a decision model that can be used by the local department of transportation to determine which type of bridge is most appropriate for a given site. Once your decision model has been developed, you will be given a scenario that will allow you to apply your model and make a recommendation for the type of bridge that should be built. Each group will present its model and defend the group's recommendation to the class and members of the community.

Driving Question: How can we develop a decision model to help us make a recommendation to the local department of transportation on the type of bridge to build for a given location?

CONTENT STANDARDS ADDRESSED IN THIS STEM ROAD MAP MODULE

A full listing with descriptions of the standards this module addresses can be found in the appendix. Listings of the particular standards addressed within lessons are provided in a table for each lesson in Chapter 4.

STEM RESEARCH NOTEBOOK

Each student should maintain a STEM Research Notebook, which will serve as a place for students to organize their work throughout this module (see p. 12 for more general discussion on setup and use of this notebook). All written work in the module should be included in the notebook, including records of students' thoughts and ideas, fictional accounts based on the concepts in the module, and records of student progress through the engineering design process. The notebooks may be maintained across subject areas, giving students the opportunity to see that although their classes may be separated during the school day, the knowledge they gain is connected.

Lessons in this module include student handouts that should be kept in the STEM Research Notebooks after completion, as well as prompts to which students should respond in their notebooks. You may also wish to have students include the STEM Research Notebook Guidelines student handout on page 26 in their notebooks.

Emphasize to students the importance of organizing all information in a Research Notebook. Explain to them that scientists and other researchers maintain detailed Research Notebooks in their work. These notebooks, which are crucial to researchers' work because they contain critical information and track the researchers' progress, are often considered legal documents for scientists who are pursuing patents or wish to provide proof of their discovery process.

STUDENT HANDOUT

STEM RESEARCH NOTEBOOK GUIDELINES

STEM professionals record their ideas, inventions, experiments, questions, observations, and other work details in notebooks so that they can use these notebooks to help them think about their projects and the problems they are trying to solve. You will each keep a STEM Research Notebook during this module that is like the notebooks that STEM professionals use. In this notebook, you will include all your work and notes about ideas you have. The notebook will help you connect your daily work with the big problem or challenge you are working to solve.

It is important that you organize your notebook entries under the following headings:

1. **Chapter Topic or Title of Problem or Challenge:** You will start a new chapter in your STEM Research Notebook for each new module. This heading is the topic or title of the big problem or challenge that your team is working to solve in this module.
2. **Date and Topic of Lesson Activity for the Day:** Each day, you will begin your daily entry by writing the date and the day's lesson topic at the top of a new page. Write the page number both on the page and in the table of contents.
3. **Information Gathered From Research:** This is information you find from outside resources such as websites or books.
4. **Information Gained From Class or Discussions With Team Members:** This information includes any notes you take in class and notes about things your team discusses. You can include drawings of your ideas here, too.
5. **New Data Collected From Investigations:** This includes data gathered from experiments, investigations, and activities in class.
6. **Documents:** These are handouts and other resources you may receive in class that will help you solve your big problem or challenge. Paste or staple these documents in your STEM Research Notebook for safekeeping and easy access later.
7. **Personal Reflections:** Here, you record your own thoughts and ideas on what you are learning.
8. **Lesson Prompts:** These are questions or statements that your teacher assigns you within each lesson to help you solve your big problem or challenge. You will respond to the prompts in your notebook.
9. **Other Items:** This section includes any other items your teacher gives you or other ideas or questions you may have.

MODULE LAUNCH

To launch the module, facilitate a class discussion about the need for bridges, including impact on a community and the types of bridges that students are familiar with. Following the discussion, the class should view a video clip related to the construction of bridges. A variety of videos can be found on the internet or on YouTube; one example is "Bridge Building Video" at *www.sciencekids.co.nz/videos/engineering/bridgebuilding.html.* After viewing the video, extend previous discussion about types of bridges, but now begin a conversation about the pros and cons of bridge types and the need to make a decision about the type of bridge each time a new bridge is planned.

Tell students that as part of their challenge in this module, they will help the local department of transportation develop a decision model to help the department decide on the best type of bridge to put in place based on the location.

PREREQUISITE SKILLS FOR THE MODULE

Students enter this module with a wide range of preexisting skills, information, and knowledge. Table 3.1 (p. 28) provides an overview of prerequisite skills and knowledge that students are expected to apply in this module, along with examples of how they apply this knowledge throughout the module. Differentiation strategies are also provided for students who may need additional support in acquiring or applying this knowledge.

Table 3.1. Prerequisite Key Knowledge and Examples of Applications and Differentiation Strategies

Prerequisite Key Knowledge	Application of Knowledge by Students	Differentiation for Students Needing Knowledge
• Apply the notion of scale factor and proportional reasoning in real-world contexts. • Graph points in the *x-y* coordinate plane and use the plot of these points to analyze data. • Generate and solve linear equations in a real-world context. • Know and be able to apply the Pythagorean theorem.	Scale Factor: • Develop a scale drawing and construct a 3-D model of a bridge in their community. Graphing: • Throughout module, collect data and display findings on a coordinate plane. Linear Equations: • From investigations, organize data and write/solve linear models to make predictions that will inform decision model. Pythagorean Theorem: • Use the Pythagorean theorem to find the length of support cables in a cable-stayed bridge.	• Do a short activation lesson for all students on scaling. • Have students work in project groups; students needing support with the concept of scaling can be grouped with students who demonstrate an understanding of the concept. • Supply students with a graphing utility.
• Have basic internet research skills. • Conduct internet research, including determining important information and reliable sources. • Have a basic understanding of figurative language, including metaphors. • Be familiar with nonfiction text structures and features and able to use them in writing.	• Use computers and the internet to research the types and uses of minerals found in your state. • Research bridges that are or have been significant in our cultural experiences as well as the various metaphorical uses of *bridge.* • Articulate the significance of *bridge* as a metaphor and use that information to understand literature.	• Provide a class guide for internet search engines. • Hold a classroom discussion about how to effectively use Boolean search terms. • Provide students the opportunity to practice assessing the credibility of various websites. • Select varied types and forms of literature and allow choice to support access for all learners. For struggling readers, reduce concept load by selecting literature that addresses familiar contexts.

POTENTIAL STEM MISCONCEPTIONS

Students enter the classroom with a wide variety of prior knowledge and ideas, so it is important to be alert to misconceptions, or inappropriate understandings of foundational knowledge. These misconceptions can be classified as one of several types: "preconceived notions," opinions based on popular beliefs or understandings; "nonscientific beliefs," knowledge students have gained about science from sources outside the scientific community; "conceptual misunderstandings," incorrect conceptual models based on incomplete understanding of concepts; "vernacular misconceptions," misunderstandings of words based on their common use versus their scientific use; and "factual misconceptions," incorrect or imprecise knowledge learned in early life that remains unchallenged (NRC 1997, p. 28). Misconceptions must be addressed and dismantled in order for students to reconstruct their knowledge, and therefore teachers should be prepared to take the following steps:

- *Identify students' misconceptions.*
- *Provide a forum for students to confront their misconceptions.*
- *Help students reconstruct and internalize their knowledge, based on scientific models. (NRC 1997, p. 29)*

Keeley and Harrington (2010) recommend using diagnostic tools such as probes and formative assessment to identify and confront student misconceptions and begin the process of reconstructing student knowledge. Keeley's *Uncovering Student Ideas in Science* series contains probes targeted toward uncovering student misconceptions in a variety of areas and may be useful resources for addressing student misconceptions in this module.

Some commonly held misconceptions specific to lesson content are provided with each lesson so that you can be alert for student misunderstanding of the science concepts presented and used during this module. The American Association for the Advancement of Science has also identified misconceptions that students frequently hold regarding various science concepts (see the links at *http://assessment.aaas.org/topics*).

SRL PROCESS COMPONENTS

Table 3.2 illustrates some of the activities in the Improving Bridge Design module and how they align to the SRL processes before, during, and after learning.

Table 3.2. SRL Process Components

Learning Process Components	Examples From Improving Bridge Design Module	Lesson Number and Learning Component
	BEFORE LEARNING	
Motivates students	Students are challenged to become experts in bridge building so that they can help the community. The students are motivated by watching a bridge collapse video.	Lesson 1, Introductory Activity/Engagement
Evokes prior learning	Students tap into their prior experience with bridges by exploring bridges in their local community.	Lesson 1, Activity/ Exploration
	DURING LEARNING	
Focuses on important features	Students brainstorm in small groups on what they know about bridges and what they still need to know. These thoughts are shared with the class and the entire class hones the list to the most important.	Lesson 2, Introductory Activity/Engagement
Helps students monitor their progress	While students are gathering data on the span length constraints of a beam bridge, the teacher chooses a group to display its data in graphing software to the class. Groups check their processes according to this model.	Lesson 2, Activity/ Exploration
	AFTER LEARNING	
Evaluates learning	In the final challenge, students create a decision model and present it to peers, members of the local department of transportation, and other members of the community for feedback.	Lesson 6, Elaboration/ Application of Knowledge
Takes account of what worked and what did not work	In the final challenge, students reflect on the review and reactions from peers and community members for their decision model.	Lesson 6, Elaboration/ Application of Knowledge

STRATEGIES FOR DIFFERENTIATING INSTRUCTION WITHIN THIS MODULE

For the purposes of this curriculum module, differentiated instruction is conceptualized as a way to tailor instruction—including process, content, and product—to various student needs in your class. A number of differentiation strategies are integrated into lessons across the module. The problem- and project-based learning approach used in the lessons is designed to address students' multiple intelligences by providing a variety of entry points and methods to investigate the key concepts in the module (for example, investigating bridges from the perspectives of science and social issues via scientific inquiry, literature, journaling, and collaborative design). Differentiation strategies for students needing support in prerequisite knowledge can be found in Table 3.1 (p. 28). You are encouraged to use information gained about student prior knowledge during introductory activities and discussions to inform your instructional differentiation. Strategies incorporated into this lesson include flexible grouping, varied environmental learning contexts, assessments, compacting, and tiered assignments and scaffolding.

Flexible Grouping. Students work collaboratively in a variety of activities throughout this module. Grouping strategies you might employ include student-led grouping, grouping students according to ability level, grouping students randomly, grouping them so that students in each group have complementary strengths (for instance, one student might be strong in mathematics, another in art, and another in writing), or grouping students according to common interests.

Varied Environmental Learning Contexts. Students have the opportunity to learn in various contexts throughout the module, including alone, in groups, in quiet reading and research-oriented activities, and in active learning through inquiry and design activities. In addition, students learn in a variety of ways, including through doing inquiry activities, journaling, reading fiction and nonfiction texts, watching videos, participating in class discussion, and conducting web-based research.

Assessments. Students are assessed in a variety of ways throughout the module, including individual and collaborative formative and summative assessments. Students have the opportunity to produce work via written text, oral and media presentations, and modeling. You may choose to provide students with additional choices of media for their products (for example, PowerPoint presentations, posters, or student-created websites or blogs).

Compacting. Based on student prior knowledge, you may wish to adjust instructional activities for students who exhibit prior mastery of a learning objective. For instance, in Lesson 4 the teacher is prompted to provide a mini lesson on the Pythagorean theorem. The use of this theorem is needed to aid the students in their exploration of cable-stayed bridges. However, if some students exhibit mastery of the application of the Pythagorean theorem, you may wish to use this time instead to introduce ELA or social studies connections with associated activities.

Tiered Assignments and Scaffolding. Based on your awareness of student ability, understanding of concepts, and mastery of skills, you may wish to provide students with variations on activities by adding complexity to assignments or providing more or fewer learning supports for activities throughout the module. For instance, some students may need additional support in identifying key search words and phrases for web-based research or may benefit from cloze sentence handouts to enhance vocabulary understanding. Other students may benefit from expanded reading selections and additional reflective writing or from working with manipulatives and other visual representations of mathematical concepts. You may also work with your school librarian to compile a set of topical resources at a variety of reading levels.

STRATEGIES FOR ENGLISH LANGUAGE LEARNERS

Students who are developing proficiency in English language skills require additional supports to simultaneously learn academic content and the specialized language associated with specific content areas. WIDA (2012) has created a framework for providing support to these students and makes available rubrics and guidance on differentiating instructional materials for English language learners (ELLs). In particular, ELL students may benefit from additional sensory supports such as images, physical modeling, and graphic representations of module content, as well as interactive support through collaborative work. This module incorporates a variety of sensory supports and offers ongoing opportunities for ELL students to work collaboratively. The focus in this module on bridges affords opportunities to access the culturally diverse experiences of ELL students in the classroom.

In differentiating instruction for ELL students, you should carefully consider the needs of these students as you introduce and use academic language in various language domains (listening, speaking, reading, and writing) throughout this module. To adequately differentiate instruction for ELL students, you should have an understanding of the proficiency level of each student. The following five overarching WIDA learning standards are relevant to this module:

- Standard 1: Social and Instructional language. Focus on social behavior in group work and class discussions.
- Standard 2: The language of Language Arts. Focus on forms of print, elements of text, picture books, comprehension strategies, main ideas and details, persuasive language, creation of informational text, and editing and revision.
- Standard 3: The language of Mathematics. Focus on numbers and operations, patterns, number sense, measurement, and strategies for problem solving.
- Standard 4: The language of Science. Focus on safety practices, scientific process, and scientific inquiry.

- Standard 5: The language of Social Studies. Focus on historical events and people, resources, geography, and environmental issues.

SAFETY CONSIDERATIONS FOR THE ACTIVITIES IN THIS MODULE

Student safety is a primary consideration in all subjects where students may interact with tools and materials with which they are unfamiliar and which may pose additional safety risks. You should ensure that your classroom set-up is in accord with your school's safety policies and that students are familiar with basic safety procedures, the location of protective equipment (e.g., safety glasses, gloves), and emergency exit procedures. For more general safety guidelines, see the Safety in STEM section in Chapter 2 (p. 18).

Internet safety is also important. You should develop an internet/blog protocol with students if guidelines are not already in place. Since students will use the internet for their research to acquire the needed data, you should monitor students' access to ensure that they are accessing only websites that you have clearly identified. Further, you should inform parents or guardians that students will create online multimedia presentations of their research and that you will closely monitor these projects. It is recommended that you not allow any website posts created by students to go public without first approving them. During this module, students will be asked to explore a bridge in their community. You should ensure that students have the appropriate parental or adult supervision when exploring their desired bridge.

DESIRED OUTCOMES AND MONITORING SUCCESS

The desired outcome for this module is outlined in Table 3.3, along with suggested ways to gather evidence to monitor student success. For more specific details on desired outcomes, see the Established Goals and Objectives section for the module (p. 23) and for the individual lessons.

Table 3.3. Desired Outcome and Evidence of Success in Achieving Identified Outcome

	Evidence of Success	
Desired Outcome	**Performance Tasks**	**Other Measures**
Students create and present a decision model that illustrates their understanding of bridge design, structure, and function.	Students are assessed on their written proposal and poster presentation of their decision model and its application to determine the appropriate bridge for a given local site(s).	Students are assessed on • how well they work together in their groups, • participation in classroom discussion, and • individual investigation activity sheets throughout module.

ASSESSMENT PLAN OVERVIEW AND MAP

Table 3.4 provides an overview of the major group and individual *products* and *deliverables,* or things that constitute the assessment for this module. See Table 3.5 for a full assessment map of formative and summative assessments in this module.

Table 3.4. Major Products and Deliverables in Lead Disciplines for Groups and Individuals

Lesson	Major Group Products and Deliverables	Major Individual Products and Deliverables
1	Short presentations about a bridge in the local community	• STEM Research Notebook entries • Individual investigation activity sheets throughout module
2	Beam bridge scale drawing and 3-D model	• STEM Research Notebook entries • Individual investigation activity sheets throughout module
3	Arch bridge scale drawing and 3-D model	• STEM Research Notebook entries • Individual investigation activity sheets throughout module
4	Suspension bridge scale drawing and 3-D model	• STEM Research Notebook entries • Individual investigation activity sheets throughout module
5	Bridge cost equation and graph	• STEM Research Notebook entries • Individual investigation activity sheets throughout module
6	Written proposal and poster presentation of decision model and its application to local sites	• STEM Research Notebook entries • Individual investigation activity sheets throughout module • Collaboration Rubric

Table 3.5. Assessment Map for Improving Bridge Design Module

Lesson	Assessment	Group/ Individual	Formative/ Summative	Lesson Objective Assessed
1	Bridge presentations	Group	Formative	• Explore the current state of infrastructure in the United States. • Explore bridge collapses in the community and describe potential causes of bridge collapses.
1	Scaling Bridges *checklist*	Group	Formative	• Using scale factor, draw and construct scale models of a bridge.
1	STEM Research Notebook *prompts*	Individual	Formative	• Research the types and uses of different minerals present in the state.
2	Beam Bridge *cluster web*	Individual	Formative	• Describe the historical impact of bridges.
2	Rock Observation *rubric*	Group	Formative	• Collect and organize data through experimentation. • Identify differences and similarities between sedimentary, igneous, and metamorphic rocks.
2	Learning activity responses (Beam Bridge Penny Challenge, Beam Bridges—Effect of Span Length, Other Beam Bridge Facts *handouts*)	Individual	Formative	• Interpret data and write a linear equation that best fits the data. • Use a linear model to solve problems in a real-world context.

Continued

Table 3.5. *(continued)*

Lesson	Assessment	Group/ Individual	Formative/ Summative	Lesson Objective Assessed
2	STEM Research Notebook *prompts*	Individual	Formative	• Identify differences and similarities between sedimentary, igneous, and metamorphic rocks.
3	Game for elementary students	Group	Formative	• Develop a game that accurately teaches the rock cycle to elementary-age children in the grade 3–5 range.
3	Arches in History Poster *checklist*	Group	Formative	• Understand the role that arches have played in the development of infrastructure across time and culture.
3	Learning activity responses (Arch Bridge Weight Test, Arch Bridge Basics, Arch Bridge—Span Length *handouts*)	Individual	Formative	• Interpret data and write a linear equation that best fits the data. • Use a linear model to solve problems in a real-world context.
3	STEM Research Notebook *prompts*	Group	Formative	• Understand the role that arches have played in the development of infrastructure across time and culture.
4	Bridges: Compare and Contrast Matrix *handout*	Individual	Summative	• Compare and contrast various bridge types.

Continued

Table 3.5. *(continued)*

Lesson	Assessment	Group/ Individual	Formative/ Summative	Lesson Objective Assessed
4	Learning activity responses (Suspension Bridge Weight Test, Suspension Bridge Basics, Cable-Stayed Bridge Basics, Cable-Stayed Bridge Investigation *handouts*)	Individual	Formative	• Interpret data and write a linear equation that best fits the data. • Use a linear model to solve problems in a real-world context. • Use the Pythagorean theorem to solve real-world problems. • Understand the strengths and limitations of suspension and cable-stayed bridges.
4	STEM Research Notebook *prompts*	Individual	Formative	• Understand the strengths and limitations of suspension and cable-stayed bridges.
5	Cost of Bridges Investigation *handouts*	Group/ Individual	Formative	• Write an equation for total cost given initial cost and cost of yearly maintenance. • Graph cost equation on a coordinate plane and describe what the y-intercept and slope mean. • Compare cost functions to determine which is the cheapest for a given time period.
5	STEM Research Notebook *prompts*	Individual	Formative	• Understand the importance of geology to roads and bridges.

Continued

Table 3.5. (*continued*)

Lesson	Assessment	Group/ Individual	Formative/ Summative	Lesson Objective Assessed
6	Proposal, poster, presentation (Written Proposal and Poster and Presentation *rubrics*)	Group	Summative	• Develop a decision model. • Use a decision model to select a bridge design for a given scenario.
6	Works Progress Administration (WPA) debate (Social Studies Debate *rubric*)	Group	Summative	• Defend a position on whether another WPA should be established.

MODULE TIMELINE

Tables 3.6–3.10 (pp. 39–40) provide lesson timelines for each week of the module. These timelines are provided for general guidance only and are based on class times of approximately 45 minutes.

Table 3.6. STEM Road Map Module Schedule for Week One

Day 1	Day 2	Day 3	Day 4	Day 5
Lesson 1 *Bridges in the Community* • Launch the module by introducing the challenge and showing the bridge-building video. • Following this, show the bridge collapse video and explore bridge infrastructure in the United States.	*Lesson 1* *Bridges in the Community* • Explore bridges in the local community. • Conduct bridge scavenger hunt and research. • Students work on researching their local bridges (e.g., age, size, folklore).	*Lesson 1* *Bridges in the Community* • Students present research on their bridges to the class.	*Lesson 1* *Bridges in the Community* • Students work on scale drawing and constructing a scale model of one bridge from their presentation.	*Lesson 1* *Bridges in the Community* • Students finish and present scale drawing and model of one bridge from their presentation.

Table 3.7. STEM Road Map Module Schedule for Week Two

Day 6	Day 7	Day 8	Day 9	Day 10
Lesson 2 *Beam Bridges* • Explore the design, structure, and function of beam bridges.	*Lesson 2* *Beam Bridges* • Continue to explore the design, structure, and function of beam bridges.	*Lesson 2* *Beam Bridges* • Explore effect span length has on a beam bridge.	*Lesson 3* *Arch Bridges* • Explore the strength of arch bridges.	*Lesson 3* *Arch Bridges* • Finish and discuss exploration of the strength of arch bridges; discuss forces involved with arch bridges.

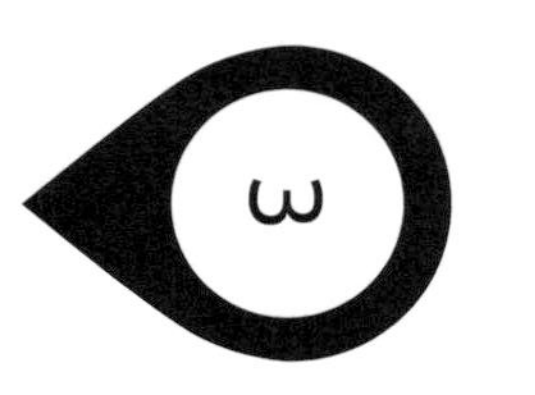

Table 3.8. STEM Road Map Module Schedule for Week Three

Day 11	Day 12	Day 13	Day 14	Day 15
Lesson 3 *Arch Bridges* • Explore the effect span length has on an arch bridge.	*Lesson 4* *Suspension and Cable-Stayed Bridges* • Explore the design, structure, and function of suspension bridges.	*Lesson 4* *Suspension and Cable-Stayed Bridges* • Continue to explore the design, structure, and function of suspension bridges.	*Lesson 4* *Suspension and Cable-Stayed Bridges* • Explore the design, structure, and function of cable-stayed bridges.	*Lesson 5* *Economics and Bridges* • Identify the costs related to constructing and maintaining a bridge.

Table 3.9. STEM Road Map Module Schedule for Week Four

Day 16	Day 17	Day 18	Day 19	Day 20
Lesson 5 *Economics and Bridges* • Continue exploring the economics of bridges.	*Lesson 6* *Putting It All Together—Decision Models* • Review the module challenge. • Begin research on what type of bridge is the appropriate choice for a given site.	*Lesson 6* *Putting It All Together—Decision Models* • Students research the four types of bridges and begin to think about what makes them the appropriate choice for a given site.	*Lesson 6* *Putting It All Together—Decision Models* • Students complete research on appropriate bridge for a given location.	*Lesson 6* *Putting It All Together—Decision Models* • Have a speaker from the local department of transportation.

Table 3.10. STEM Road Map Module Schedule for Week Five

Day 21	Day 22	Day 23	Day 24	Day 25
Lesson 6 *Putting It All Together—Decision Models* • Students finalize bridge decision model and apply it to their local sites.	*Lesson 6* *Putting It All Together—Decision Models* • Students prepare presentations.	*Lesson 6* *Putting It All Together—Decision Models* • Students prepare presentations.	*Lesson 6* *Putting It All Together—Decision Models* • Students present decision models.	*Lesson 6* *Putting It All Together—Decision Models* • Students present decision models.

RESOURCES

Teachers have the option to coteach portions of this module and may want to combine classes for activities such as mathematical modeling, geometric investigations, discussing social influences, or conducting research. The media specialist can help teachers locate resources for students to view and read about bridges and related engineering content. Special educators and reading specialists can help find supplemental sources for students needing extra support in reading and writing. Additional resources may be found online. Community resources for this module may include civil engineers or department of transportation representatives.

REFERENCES

Johnson, C. C., T. J. Moore, J. Utley, J. Breiner, S. R. Burton, E. E. Peter-Burton, J. Walton, and C. L. Parton. 2015. The STEM road map for grades 6–8. In *STEM road map: A framework for integrated STEM education,* ed. C. C. Johnson, E. E. Peters-Burton, & T. J. Moore, 96–123. New York: Routledge. *www.routledge.com/products/9781138804234.*

Keeley, P., and R. Harrington. 2010. *Uncovering student ideas in physical science, volume 1: 45 new force and motion assessment probes.* Arlington, VA: NSTA Press.

National Research Council (NRC). 1997. *Science teaching reconsidered: A handbook.* Washington, DC: National Academies Press.

WIDA. 2012. 2012 amplification of the English language development standards: Kindergarten–grade 12. *www.wida.us/standards/eld.aspx.*

IMPROVING BRIDGE DESIGN LESSON PLANS

John Weaver, Toni A. Ivey, Juliana Utley, Adrienne Redmond-Sanogo, Sue Christian Parsons, Janet B. Walton, Carla C. Johnson, and Erin Peters-Burton

Lesson Plan 1: Bridges in the Community

In this lesson, students have the opportunity to explore the state of infrastructure in the United States. Using bridges as a lens, students are presented with the challenge of developing a decision model for the local department of transportation. In the process of developing their models, students explore properties and attributes of various bridges to ensure their model helps city planners and civil engineers make appropriate choices when it comes to function, design, and cost of a bridge.

ESSENTIAL QUESTIONS

- What is the current state of infrastructure in the United States?
- How do the decisions of the past affect us today?
- How do the decisions we make today affect our future?
- How do engineers use scaling to test and explore bridges?
- What role do minerals play in the construction of roads?
- How do bridges affect society?
- What are the economic and social impacts of a bridge collapse?
- What are some reasons bridges fail?

ESTABLISHED GOALS AND OBJECTIVES

At the conclusion of this lesson, students will be able to do the following:

- Explore the current state of infrastructure in the United States.
- Use scale factor to draw and construct scale models of a bridge.

- Explore bridge collapses in their community and describe potential causes of bridge collapses.
- Research the types and uses of different minerals present in their state.
- Discuss the economic and social implications of a bridge collapse in their community.
- Construct knowledge about bridges, including their significance in our culture and resulting presence in literature.
- Use and evaluate a variety of sources to find and select information about bridges, citing appropriately.
- Develop understandings about the significance of place (setting) in lives and literature.
- Understand and recognize symbol and metaphor (specifically, the symbolic significance of bridges).
- Identify the criteria for a substance to be considered a mineral.
- Research to learn which minerals naturally occur in their state.
- Identify different uses for minerals in road and bridge construction.

TIME REQUIRED

- 5 days (approximately 45 minutes each day; see Table 3.6, p. 39)

MATERIALS

- STEM Research Notebooks (1 per student; see p. 26 for STEM Research Notebook Guidelines student handout)
- Internet access for research
- Handouts (attached at the end of this lesson)
- Various materials to build bridges:
  - Craft sticks
  - Toothpicks
  - Clear tape
  - Glue
  - Card stock

  - Safety glasses or safety goggles
  - Rulers
- Group materials (one per group of students):
  - Graduated cylinder
  - Balance/scale
  - Water
  - Penny
  - Nail
  - Glass plate
  - Streak plate
  - A set of common minerals for identification per group (such as calcite, gypsum, halite, talc, fluorite, quartz, graphite, galena, magnetite, biotite, muscovite, hematite, and pyrite)

SAFETY NOTES

1. All laboratory occupants must wear safety glasses or goggles during all phases of this inquiry activity.
2. Use caution when working with sharps (e.g., sticks, toothpicks, glass plate) to avoid cutting or puncturing skin.
3. Make sure all materials are put away after completing the activity.
4. Wash hands with soap and water after completing this activity.

CONTENT STANDARDS AND KEY VOCABULARY

Table 4.1 (p. 46) lists the content standards from the *Next Generation Science Standards (NGSS)*, *Common Core State Standards (CCSS)*, and the Framework for 21st Century Learning that this lesson addresses, and Table 4.2 (p. 48) presents the key vocabulary. Vocabulary terms are provided for both teacher and student use. Teachers may choose to introduce some or all of the terms to students.

Table 4.1. Content Standards Addressed in STEM Road Map Module Lesson 1

NEXT GENERATION SCIENCE STANDARDS

PERFORMANCE EXPECTATIONS

- MS-ESS2-1. Develop a model to describe the cycling of Earth's materials and the flow of energy that drives this process.

SCIENCE AND ENGINEERING PRACTICES

Analyzing and Interpreting Data

Analyzing data in 6–8 builds on K–5 experiences and progresses to extending quantitative analysis to investigations, distinguishing between correlation and causation, and basic statistical techniques of data and error analysis.

- Analyze and interpret data to provide evidence for phenomena.

Constructing Explanations and Designing Solutions

Constructing explanations and designing solutions in 6–8 builds on K–5 experiences and progresses to include constructing explanations and designing solutions supported by multiple sources of evidence consistent with scientific ideas, principles, and theories.

- Construct a scientific explanation based on valid and reliable evidence obtained from sources (including the students' own experiments) and the assumption that theories and laws that describe the natural world operate today as they did in the past and will continue to do so in the future.

DISCIPLINARY CORE IDEA

ESS3.A: Natural Resources

- Humans depend on Earth's land, ocean, atmosphere, and biosphere for many different resources. Minerals, fresh water, and biosphere resources are limited, and many are not renewable or replaceable over human lifetimes. These resources are distributed unevenly around the planet as a result of past geologic processes.

CROSSCUTTING CONCEPT

Influence of Science, Engineering, and Technology on Society and the Natural World

- All human activity draws on natural resources and has both short and long-term consequences, positive as well as negative, for the health of people and the natural environment.

Continued

Table 4.1. *(continued)*

COMMON CORE STATE STANDARDS FOR MATHEMATICS

MATHEMATICAL PRACTICES

- MP1. Make sense of problems and persevere in solving them.
- MP2. Reason abstractly and quantitatively.
- MP3, Construct viable arguments and critique the reasoning of others.
- MP6. Attend to precision.

COMMON CORE STATE STANDARDS FOR ENGLISH LANGUAGE ARTS

READING STANDARDS

- RL.8.1. Cite the textual evidence that most strongly supports an analysis of what the text says explicitly as well as inferences drawn from the text.
- RL.8.2. Determine a theme or central idea of a text and analyze its development over the course of the text, including its relationship to the characters, setting, and plot; provide an objective summary of the text.

WRITING STANDARDS

- W.8.6. Use technology, including the internet, to produce and publish writing and present the relationships between information and ideas efficiently as well as to interact and collaborate with others.
- W.8.7. Conduct short research projects to answer a question (including a self-generated question), drawing on several sources and generating additional related, focused questions that allow for multiple avenues of exploration.
- W.8.8. Gather relevant information from multiple print and digital sources, using search terms effectively; assess the credibility and accuracy of each source; and quote or paraphrase the data and conclusions of others while avoiding plagiarism and following a standard format for citation.

SPEAKING AND LISTENING STANDARDS

- SL.8.1. Engage effectively in a range of collaborative discussions (one-on-one, in groups, and teacher-led) with diverse partners on grade 8 topics, texts, and issues, building on others' ideas and expressing their own clearly.
- SL.8.1.A. Come to discussions prepared, having read or researched material under study; explicitly draw on that preparation by referring to evidence on the topic, text, or issue to probe and reflect on ideas under discussion.
- SL.8.1.B. Follow rules for collegial discussions and decision making, track progress toward specific goals and deadlines, and define individual roles as needed.
- SL.8.1.C. Pose questions that connect the ideas of several speakers and respond to others' questions and comments with relevant evidence, observations, and ideas.

Continued

Table 4.1. *(continued)*

SPEAKING AND LISTENING STANDARDS *(continued)*
• SL.8.1.D. Acknowledge new information expressed by others, and, when warranted, qualify or justify their own views in light of the evidence presented. • SL.8.4. Present claims and findings, emphasizing salient points in a focused, coherent manner with relevant evidence, sound valid reasoning, and well-chosen details; use appropriate eye contact, adequate volume, and clear pronunciation.
LANGUAGE STANDARD
• L.8.5. Demonstrate understanding of figurative language, word relationships, and nuances in word meanings.
FRAMEWORK FOR 21ST CENTURY LEARNING
• Interdisciplinary Themes: Global Awareness; Environmental Literacy; Civic Literacy • Learning and Innovation Skills: Creativity and Innovation; Critical Thinking and Problem Solving; Communication and Collaboration; Information Literacy • Information, Media, and Technology Skills: Media Literacy; ICT Literacy • Life and Career Skills: Flexibility and Adaptability; Initiative and Self-Direction; Social and Cross-Cultural Skills; Productivity and Accountability; Leadership and Responsibility

Table 4.2. Key Vocabulary for Lesson 1

Key Vocabulary	Definition
city planner	a person who works to ensure that a city has a cohesive plan for development and infrastructure; evaluates infrastructure, assesses needs, and analyzes impact of infrastructure changes on the community to create a strategy to guide a city's growth
civil engineer	an engineer who designs and builds roads, bridges, and other public works
cleavage	the way a mineral splits or cleaves
color	the color of a mineral
density	the amount of mass per unit of volume
hardness	how hard a mineral is; hardness refers to how much a mineral resists scratch and is measured with the Mohs hardness scale
infrastructure	"the basic physical and organizational structures and facilities (e.g., buildings, roads, and power supplies) needed for the operation of a society or enterprise" (Oxford Dictionaries 2015)
luster	how well a mineral reflects light

Table 4.2. *(continued)*

Key Vocabulary	Definition
mineral	a naturally occurring, inorganic (not produced from living organisms), solid substance that has a consistent chemical formula and structure
Mohs hardness scale	used to determine the hardness of mineral on a scale of 1 to 10 with harder minerals having a higher number on the scale
physical property	a physical characteristic of a substance that describes its matter (e.g., color, density) but not how it reacts chemically with other substances
rock	any naturally occurring substance generally composed of one or more minerals
scale factor	"the ratio of any two corresponding lengths in two similar geometric figures" (Simmons 2014)

TEACHER BACKGROUND INFORMATION

Mathematics

As part of this lesson, students create scaled models and drawings that represent an actual bridge students have explored. Scale factor is addressed in previous grade levels; however, students may need additional support. The toy car activity is included to help activate prior knowledge of scaling. More information on scaling and scale factor can be found at the following websites:

- *http://illuminations.nctm.org/Activity.aspx?id=4207*
- *http://study.com/academy/lesson/what-is-a-scale-factor-definition-formula-examples.html*

Science

What Are Minerals? Scientists consider a substance a mineral if it meets the following five criteria: (1) naturally occurring—not produced by people, (2) inorganic—not made by an organism, (3) solid—not a liquid or gas, (4) definite chemical composition, and (5) ordered internal structure.

The phrase *definite chemical composition* means that anytime someone encounters a mineral, it will always have the same chemical formula. For instance, the mineral halite (which is rock salt and table salt) has a chemical composition of NaCl, sodium chloride. The phrase *ordered internal structure* means that the atoms of the mineral are arranged in a pattern that repeats and is systematic. In other words, for halite, the sodium and chlorine atoms always connect to each other in the same way. This property can be useful for mineral identification as some minerals have a very distinctive shape. However, the shape of a mineral can also be misleading and should be used with caution as an identifier.

Properties of Minerals. All minerals have a set of unique properties that help geologists identify them in the field. Physical properties of minerals include color, streak, hardness, luster, cleavage, fracture, magnetism, and solubility. The Mineral Identification Worksheet at the end of this lesson contains definitions for the various properties of minerals that students will identify.

Minerals and Infrastructure. Minerals play a critical role in the world economy. The development and maintenance of basic infrastructure, including pipes, roads, and bridges, depends on an adequate supply of raw Earth materials. The raw materials that are required for most construction projects are ultimately derived from Earth materials that are mined or dredged from the sea; this includes almost all concrete, asphalt, bricks, clay, cement, and ceramics. Metals used in construction must also be mined.

More information on minerals can be found at the following websites:

- *www.bgs.ac.uk/downloads/browse.cfm?sec=12&cat=120*
- *www.geology.com/minerals/what-is-a-mineral.shtml*

Social Studies

You should be familiar with the fascinating history of the Brooklyn Bridge as well as other bridges that play a central role in American history and culture. Lynn Curlee's book *Brooklyn Bridge* (Atheneum Books for Young Readers, 2001) used in the lesson is a good place to start. To explore deeper, consider reading *The Great Bridge: The Epic Story of the Building of the Brooklyn Bridge* by David McCullough (Simon & Schuster, 1983) and *The Brooklyn Bridge: The Story of the World's Most Famous Bridge and the Remarkable Family That Built It* by Elizabeth Mann (Mikaya Press, 2006). Additional background information on the Brooklyn Bridge can be found at *www.history.com/topics/brooklyn-bridge.*

A good place to begin identifying other significant bridges is the Historic Bridge Foundation's website: *http://historicbridgefoundation.com/bridges-in-the-u-s.* The Fun Times travel site, though not a scholarly source, also provides a good list of famous bridges that offers a starting point for exploration: *http://travel.thefuntimesguide.com/2014/01/famous-bridges.php.*

COMMON MISCONCEPTIONS

Students will have various types of prior knowledge about the concepts introduced in this lesson. Table 4.3 outlines some common misconceptions students may have concerning these concepts. Because of the breadth of students' experiences, it is not possible to anticipate every misconception that students may bring as they approach this lesson. Incorrect or inaccurate prior understanding of concepts can influence student learning in the future, however, so it is important to be alert to misconceptions such as those presented in the table.

Table 4.3. Common Misconceptions About the Concepts in Lesson 1

Topic	Student Misconception	Explanation
Scaling	"To scale" means to make something larger.	Scale factor used may be greater than 1 if the depiction is larger than the item being represented (e.g., cell representation) or less than 1, such as a bridge.
Models	Models are only 3-D.	Models in fact can be 2-D representations such as pictures, graphs, written descriptions, blueprints, and other representations, as well as a 3-D model.
	A model's usefulness is based solely on the model's physical resemblance to the object being modeled.	A physical model is a smaller or larger physical copy of an object such as a bridge. The model represents a similar object in the sense that scale is an important characteristic of the model.
Force (a push or pull between objects)	Nonmoving objects do not exert a force (e.g., a tabletop or roadway).	Stationary objects can exert forces on other objects. For example, when you roll a toy car over a tabletop, there is a frictional force between the table and the car wheels.
Weathering, erosion, and deposition (the breaking down of a material, the movement of the material to another area)	Water and wind cannot wear away rock and other building materials.	Water and wind can wear away rock and other building materials.
	Water freezing in cracks cannot break apart rock.	Water freezing in cracks can break apart rock.
Earth materials (types and uses of rocks)	Rocks have little practical use.	Rocks are used for a wide variety of consumer and industrial purposes. For example, rocks are used in paper production, cement, household cleaners, jewelry, pencils, chalk, glass, and building materials such as bricks and kitchen countertops. See the Minerals Education Coalition web page for more examples of how rocks are used (*www.mineralseducationcoalition.org/mining-minerals-information*).
	All rocks are the same.	There are three major types of rocks (sedimentary, igneous, and metamorphic), which have a variety of mineral compositions.

PREPARATION FOR LESSON 1

Mathematics

Prior to beginning this lesson, identify a variety of bridges in the community that students will be able to visit outside of school time. You should take a picture of each bridge and note its location. Contact the local department of transportation to gain resources that students can use to research the bridge assigned to them. You should visit each site and assess for hazards before assigning locations to students.

One week before beginning this lesson, give each student a copy of the Bridge Scavenger Hunt handout and divide students into teams. Create a letter to send home to parents so they are aware of the assignment and can provide supervision when the students visit their assigned bridge. Have parents sign and return the letter. Keep this on file during the school year. This letter should include the address of the bridges.

A computer lab or a classroom set of web-enabled devices will need to be reserved so students are able to research the bridges they have found.

Science

Prepare a tray of minerals for students to identify. You will need one tray of minerals per each student group. The minerals should be numbered and should be common across groups. Have on hand a mineral identification chart to use as a key for the mineral identification activity. One such chart can be found at *www.geology.com/minerals/mineral-identification.shtml.*

LEARNING COMPONENTS

Introductory Activity/Engagement

Connection to the Challenge: Begin each day of this lesson by directing students' attention to the driving question for the module and challenge: How can we develop a decision model to help us make a recommendation to the local department of transportation on the type of bridge to build for a given location? Ask students why bridges are so important and what impact a bridge might have on a community. Hold a brief student discussion of how their learning in the previous days' lesson(s) contributed to their ability to create their plan and build their prototype. You may wish to hold a class discussion, creating a class list of key ideas on chart paper, or you may wish to have students create a notebook entry with this information.

Driving Question for Lesson 1: What is the condition of the bridges found in our community?

Mathematics Class: Show a bridge collapse video. Examples include the "Minnesota Bridge Collapse" (*www.youtube.com/watch?v=CMdv2wRaqo4*) and the "Tacoma Narrows Bridge Collapse" *(www.youtube.com/watch?v=lXyG68_caV4)*.

Lead the class through a discussion after watching the videos by asking questions such as the following:

- What is one thing that surprised you about the collapse of these bridges?
- What do you think caused the collapse?
- How might the collapse have been prevented?
- What impact will the bridge collapse have on the local community?

Direct students to websites about bridges, such as *www.infrastructurereportcard.org/cat-item/bridges*. If students are not able to access these sites digitally, the information may be printed and passed out to groups for discussion. After students have had time to explore the sites, lead the students through a class discussion exploring what they found. Potential discussion questions include the following:

- What is infrastructure?
- What do these sites say about the state of the infrastructure in the United States?
- What kind of infrastructure do you see in your community?
- What is the condition of the bridges, utilities, roads, and other infrastructure in your town?

Student Challenge: Next, give the students the following challenge: *Because of the current state of bridges in the United States, we are going to spend the next few weeks researching, designing, testing, and constructing bridges. Our challenge is to help the local department of transportation make better choices that will have a positive impact on our nation's infrastructure. By making better decisions, we can help ensure that future bridges are sustainable and appropriate for the community in which they are built.*

As we discuss the variables involved in building a bridge, you will be working in groups to develop a decision model that can be used by the department of transportation to determine which type of bridge is most appropriate for a given site. Once your decision model has been developed, you will be given a scenario that will allow you to apply your model and make a recommendation for the type of bridge that should be built. Each group will present its model and defend its recommendation to the class and members of the community.

Science Connection: Ask the class the following questions:

- How did you get to school today? (List student responses. Most will answer by automobile, bus, bike, or foot.)
- What are those things made of? (List student responses. If students walked to school, be sure to remind them that their shoes are made of materials.)
- How could you categorize these materials? (List student responses.)

Allow the students a few minutes to work at their tables and complete an open sort, categorizing the materials they described. Once students have a categorization system ready, have them share their ideas with the class. If students do not come to it on their own, ask students which of the materials are *grown* and which are not. After a short discussion, encourage students to re-sort their materials into items that can be grown and those that are artificial (or constructed by humans). As students complete this activity, they should notice that most of the materials listed cannot be grown. Materials listed will vary by the lists students populate. Items that are grown would include leather, cotton, and other natural fabrics.

Inform the students that they are going to spend the next few weeks learning about Earth materials because they affect much of our daily life. Students should be reminded that for all of our resources, we have to either *grow it* or *mine it*: Either our resources come from plants and animals, or we have to mine for them.

STEM Research Notebook Prompt

Ask students to reflect on all of the *mined* materials that they used to get to school, such as the roads they traveled on. Remind students that these are natural Earth materials, and that it is important to gain an understanding of Earth materials, including minerals, so that they can understand how they are used in the building and maintenance of roads and bridges.

ELA Connection: Read aloud Lynn Curlee's book *Brooklyn Bridge* as an introduction to this iconic U.S. monument. Explain to students that this bridge has inspired myriad poems, stories, and other works of art. Display some of that work; one good choice would be Joseph Stella's painting *The Brooklyn Bridge: Variation on an Old Theme* along with Warren Woessner's "River Song," a poem inspired by Stella's painting. (For both works, see Jan Greenberg's *Heart to Heart: New Poems Inspired by Twentieth-Century American Art* [Abrams, 2001].) Discuss why the bridge has inspired so many works of art. What makes this bridge so fascinating?

Social Studies Connection: Pose the following scenario: *In 2007, the Minnesota Bridge along Interstate 35 collapsed, resulting in multiple deaths. What do you think will be the social* *and economic impacts of a bridge collapse like this? How do you think a collapse like this would affect our community?*

Ask students to discuss with a partner how a bridge collapse might affect a community decision to replace the bridge. Then have students share their ideas with the class, and point out how these ideas might play into the development of their decision model.

STEM Research Notebook Prompt

Ask students do a quick write to determine what the students believe would be the social and economic impact of a bridge collapse.

Activity/Exploration

Mathematics Class: Divide the class into groups of three or four students. A week prior to the beginning of the lesson, give students the Bridge Scavenger Hunt handout (p. 65). As a homework assignment, students will need to locate a bridge in their community, visit it if possible, and bring a picture of the bridge to class. In class, groups will discuss the age, materials used, folklore, size, structure, and function of these local bridges. Students will use the information they located about their assigned bridge, plus any additional information you are able to locate (e.g., information from the local department of transportation) to develop a 5-minute presentation about their bridge. Ideally, students should be given computer access to conduct their research and build their presentation. If possible, consider reserving a computer lab or instructing students to bring their own devices that have internet access. It is recommended that students use presentation software such as PowerPoint, Prezi, or Glogster to create an interesting and engaging presentation.

STEM Research Notebook Prompt

Have the students complete a 3-2-1 protocol in their notebook. Students should do the following:

- *Describe three interesting facts you learned about bridges that might help you with your decision model.*
- *List two questions you still have about bridges that could help you in developing your decision model.*
- *Name one thing you want to remember about your presentation tomorrow.*

Science Connection: Have students develop a KWL chart (A three-columned table where the *K*, *W*, and *L* stand for what they think they KNOW, what they WANT to know, and what they LEARNED). To review what students remember about minerals from earlier grades, ask the students, "What do you know about minerals?" Once students

have reflected individually under the *K* column, they will share their ideas with the class. Record student responses. Follow up with questions such as the following:

- How do you know?
- What are the criteria for something to be considered a mineral?

After a short discussion have students record things that they want to know in the *W* column and what they learned from their classmates under the *L* column.

Divide students into small groups to investigate how to identify minerals based on their physical properties. Assign students' procedural roles (e.g., materials collector, recordkeeper, facilitator, timekeeper) to help keep all students on task. In these groups, the students will determine the general characteristics, density, hardness, color, streak, and luster of different minerals (see Mineral Identification Worksheet, p. 62).

STEM Research Notebook Prompt

Pose the following question: *How might the mineral composition of an area affect the type of bridge you could build in a particular location?* Have students discuss the question in pairs or small groups. As students share their thoughts, use chart paper to capture their ideas for them to come back to as they develop their decision model. Post this chart paper on the wall in your classroom.

ELA Connection: Have students conduct internet research on the history of the Brooklyn Bridge, paying particular attention to how the building of the bridge affected human lives and changed the communities it joined. Students may conduct investigations individually and then share findings in class, or you may assign particular events for groups of students to explore and report. Review appropriate use and citation of sources.

STEM Research Notebook Prompt

Pose the following questions: *How might the influence of bridges on people's daily lives and the community affect a decision model for building a bridge? Should the impact of a bridge on human lives be a consideration in your decision model? Should your decision model have a yes/no build link?*

Social Studies Connection: Task students with conducting internet research to determine if there have been any bridge collapses in their area. If so, what were the causes and implications of the collapse? How did it change people's daily routines? What were the economic and social implications? What was done to alleviate those issues while the bridge was reconstructed? Have students share their findings with the class.

STEM Research Notebook Prompt

Have students' journal about how engineers can learn from studying the causes and implications of a bridge collapse and what role this might play in the development of their decision model.

Explanation

Mathematics Class: Students meet in their groups to finalize their presentations and present their information to the class using their desired presentation method. As the students are presenting, the rest of the class should use a 3-2-1 protocol by writing down 3 things they found interesting, 2 things they didn't know, and 1 question they have about the material in the presentation. After each group presents, help facilitate a short discussion using the 3-2-1 protocols.

STEM Research Notebook Prompt

Provide students with time to journal about what they learned from the presentations and the 3-2-1 discussions that they might find useful in developing their decision model.

Science Connection: Students report their findings from their investigations (Mineral Identification Worksheet, p. 62). To further the student thinking, explain the criteria for something to be considered a mineral by showing the class the following:

- An introductory video on minerals such as "A Brief Introduction to Minerals" (*www.youtube.com/watch?v=8a7p1NFn64s*)
- A video of a mineralogist to show why it is important to learn about minerals and what scientists hope to learn from minerals (one example is a video on mineralogist Elizabeth Arredondo: *www.pbslearningmedia.org/resource/eb6d4ef0-2ec1-4c63-b639-c38b58380105*).

STEM Research Notebook Prompt

Ask students to answer the following questions in their STEM Research Notebook individually:

- *What is the importance of understanding minerals and the role of a mineralogist in the decisions an engineer makes in constructing a bridge?*
- *Why might this knowledge help us make better decisions about what type of bridge to construct?*

Facilitate a discussion with students to help them connect the importance of understanding minerals and the role of a mineralogist in the decisions an engineer makes in

constructing a bridge and choosing the type of bridge. To aid this discussion, you could either use chart paper to record student ideas or have students keep track of other students' ideas in their STEM Research Notebook.

ELA Connection: Discuss how place plays a significant role in the life stories of individuals and communities. Similarly, you can think of places as having "life stories" themselves: histories of how they came to be and what happened to them along the way. Humans and the places they inhabit are integrally intertwined. When a place is actually something we built—conceptualized and worked to realize—those stories seem to take on particularly powerful significance to us. Some of these become "monuments," places that become placeholders for our community values. For example, ask students, "What do you think the Brooklyn Bridge might stand for in the American experience?"

STEM Research Notebook Prompt

Have students complete a quick write about either (a) bridges in their area that might have some significance similar to the Brooklyn Bridge and why or (b) what significance might this idea of a bridge as a monument in a community play in deciding on a type of bridge and its design.

Social Studies Connection: Students develop posters that display what they found while researching bridge collapses in their area. They will display them in the school foyer for their peers to peruse (gallery walk).

STEM Research Notebook Prompt

As students do a gallery walk of the posters, have them jot down ideas they found interesting and that they should keep in mind for their decision model.

Elaboration/Application of Knowledge

Mathematics Class: To help students recall their prior knowledge of scale factor, pass around a toy car (e.g., Matchbox or Hot Wheels). Many toy cars have a scale printed on the bottom of the car. As the car is being passed around the class, encourage students to make observations. Once all students have had an opportunity explore the car, facilitate a class discussion using the following questions:

- What do you know about Matchbox cars?
- What do the numbers on the bottom mean?
- What does it mean to be a scale model?
- How could we figure out how big the real car would be?
- Why would scale models be important when constructing a bridge?

STEM Research Notebook Prompt

Have students respond to following prompt: *As we begin to study bridges, it is important that we understand and have the ability to scale bridges. Bridges are enormous and very expensive. It is important that we can create and test scale models of the bridges we are exploring so that we can test at a minimal cost. Your task is to draw a scaled diagram and construct a 3-D model of your bridge. We will NOT be testing these bridges for strength. The goal of this activity is to bring one of these bridges into the classroom so we can take a closer look at it.*

Students continue to work on scaling their bridges. As students complete their scale drawings, they should begin constructing their model. It is important to stress that these bridges will not be tested for strength. The focus is constructing a scale model of the bridge. By the end of this lesson, students will have completed a scale drawing of their bridge and a scaled 3-D model of their bridge. Once students have finished their drawings and models, display them around the room and have students do a gallery walk.

STEM Research Notebook Prompt

Have students respond to the following prompts in their STEM Research Notebooks:

- *What were some similarities between the various bridges?*
- *What were some unique features of the bridges you saw?*
- *Describe three different designs you saw and write about how your understanding of bridge designs has changed.*

Science Connection: Students investigate the types of mineral resources found in their state and create a multimedia presentation that explains the following:

- The types of minerals that are found in the state
    - Where are the minerals found?
    - What are the properties of the minerals?
    - What are the minerals used for?
- Are any of these minerals used in road and bridge construction?
    - What are they used for in the construction of roads and bridges?

Connection to the Challenge: Have students describe how minerals are used in road and bridge construction and how this might be important to know for their decision model.

ELA Connection: Students explore the "life stories" of other well-known American bridges and use their research to write a "biography of a bridge." Each biography should include how the bridge came to be, significant related events in the lives of individuals and communities, and the ways the bridge has affected the community around it. Students should conduct research out of class during Lesson 2; work toward writing and publishing will commence with Lesson 3. Although research may be conducted individually, partner or small group work is recommended so that students ultimately coauthor their bridge biographies. See the rubric on page 145.

Social Studies Connection: Students look for patterns in the causes and implications of bridge collapses.

STEM Research Notebook Prompt

Have students reflect about their findings and the implications an engineer should consider when making decisions about the location and type of bridge.

Evaluation/Assessment

Students may be assessed on the following performance tasks and other measures listed.

Performance Tasks

- Bridge Presentation Rubric
- Scaling Bridges Checklist
- Multimedia Rubric—Science

Other Measures

- STEM Research Notebook Entries. You should regularly read and respond to students in their STEM Research Notebooks. Your response should not indicate whether students' entries are right or wrong. Instead, include comments or questions that will push and stretch students' thinking and can aid students in moving toward development of their decision model.
- Completion of the Mineral Identification Worksheet.

INTERNET RESOURCES

Information about scaling and scale factor

- *http://illuminations.nctm.org/Activity.aspx?id=4207*
- *http://study.com/academy/lesson/what-is-a-scale-factor-definition-formula-examples.html*

Information about minerals

- *www.bgs.ac.uk/downloads/browse.cfm?sec=12&cat=120*
- *www.geology.com/minerals/what-is-a-mineral.shtml*

Examples of how rocks are used

- *www.mineralseducationcoalition.org/mining-minerals-information*

"A Brief Introduction to Minerals" video

- *www.youtube.com/watch?v=8a7p1NFn64s*

Video about mineralogist Elizabeth Arredondo

- *www.pbslearningmedia.org/resource/eb6d4ef0-2ec1-4c63-b639-c38b58380105*

Mineral identification chart

- *www.geology.com/minerals/mineral-identification.shtml*

Brooklyn Bridge background information

- *www.history.com/topics/brooklyn-bridge*

Information about other significant bridges

- *http://historicbridgefoundation.com/bridges-in-the-u-s*
- *http://travel.thefuntimesguide.com/2014/01/famous-bridges.php*

"Minnesota Bridge Collapse" video

- *www.youtube.com/watch?v=CMdv2wRaqo4*

"Tacoma Narrows Bridge Collapse" video

- *www.youtube.com/watch?v=lXyG68_caV4*

Website about bridges

- *www.infrastructurereportcard.org/cat-item/bridges*

Name: ______________________________

STUDENT HANDOUT, PAGE 1

MINERAL IDENTIFICATION WORKSHEET

SAFETY

1. All laboratory occupants must wear safety glasses or goggles during all phases of this inquiry activity.
2. Do not lick, taste, or eat any mineral samples.
3. Report spills immediately and avoid walking in areas where water has been spilled on the floor.
4. Keep away from electrical sources when working with water because of the potential shock hazard.
5. Use caution when working with sharps (e.g., nail, glass plate) to avoid cutting or puncturing skin.
6. Make sure all materials are put away after completing the activity.
7. Wash hands with soap and water after completing this activity.

MATERIALS (PER GROUP)

- Tray of mineral samples
- Mass balance
- Graduate cylinder or beaker
- Water
- Hand lens or magnifying glass
- Glass plate
- Nail
- Penny
- Safety glasses or safety goggles

DIRECTIONS

Use the following definitions of physical properties of minerals to fill in the Mineral Identification Chart (see separate handout). After you have filled in the chart, go to your teacher for a mineral identification key to help you identify the names of your samples.

Appearance: Using your hand lens or magnifying glass, record the physical appearance and feel.

Density: Density (g/mL) is the amount of mass (g) per unit volume (mL). Using the graduate cylinder (or beaker) and the mass balance, calculate the density for each mineral sample:

Hardness: The hardness of a mineral is based on common sense and direct observation; a mineral that is harder will scratch a softer mineral. The Mohs hardness scale ranks minerals on a scale of 1 to 10; diamond is the hardest mineral and scores a 10 on the scale and talc is a very soft mineral and scores a 1 on the scale.

Name: ______________________________

STUDENT HANDOUT, PAGE 2

MINERAL IDENTIFICATION WORKSHEET

MOHS HARDNESS SCALE

1	2	3	4	5	6	7	8	9	10
Talc	Gypsum	Calcite	Fluorite	Apatite	Feldspar	Quartz	Topaz	Corundum	Diamond

To determine the hardness of a mineral, try to scratch the surface of your unknown sample with a mineral or substance that has a known hardness. If your unknown substance can be scratched by quartz (7) but cannot be scratched by feldspar (6), then it has a hardness value between 6 and 7. Pyrite, for example, has a hardness value between 6 and 6.5. You can test the hardness of minerals by using common tools with known hardness values.

HARDNESS VALUES OF COMMON TOOLS

2.5	3.5	5.5	6.5	7	8.5
Fingernail	Penny	Glass	Steel nail	Porcelain	Emery cloth

If your unknown sample can be scratched by a copper penny (3.5) but cannot be scratched by your fingernail (2.5), then the sample has a hardness between 2.5 and 3.5.

Color: Color is a mineral property that we tend to notice first. Sometimes, the color is helpful in identifying a mineral; sometimes it is not.

Streak: Test for streak by firmly rubbing a mineral sample across an unglazed tile of white porcelain (called a streak plate). A streak occurs when the rubbing of the mineral leaves behind a line of powder (called the streak). Although some minerals may come in a variety of colors, the color of the streak is always the same. For instance, amethyst (violet), rose quartz (pink), and smoky quartz (brown) are all different colors of quartz; however, they all leave behind a white powder streak. If no streak, write *none* in the Mineral Identification Chart.

Luster: Luster is the way a mineral reflects light and should be observed on a fresh, untarnished surface. Luster consists of two general categories: *metallic* and *nonmetallic.* Other terms used to describe luster include *vitreous* (e.g., quartz), *greasy* (e.g., turquoise), and *pearly* (e.g., talc).

Cleavage: If you break a sample into smaller parts, the mineral may break along clean lines; these breaks occur along planes of weakness in the crystal's structure. Some minerals, such as calcite, have "perfect" cleavage because they have clean breaks on all planes. Other minerals, such as gypsum, have poor cleavage because they do not have clean breaks on all crystal panes.

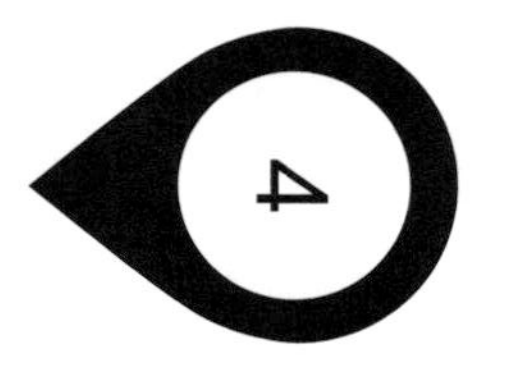

Name: ______________________

STUDENT HANDOUT

MINERAL IDENTIFICATION CHART

Sample Number	Appearance	Density	Hardness	Color	Cleavage	Streak	Luster	Mineral Name

Name: ____

STUDENT HANDOUT

BRIDGE SCAVENGER HUNT

Because of the current state of bridges in the United States, we are going to spend the next few weeks researching, designing, testing, and constructing bridges. To learn more about bridges, your group will investigate bridges in your community. With adult supervision, you will visit and explore one local bridge. (Your teacher will assign your group to a specific bridge.) Over the next week, each member of your team will visit the bridge (either individually or as a group). When visiting the bridge, you need to do the following:

- Take a picture of your bridge.
- Make observational notes about your bridge:
  - Building materials used in bridge construction
  - Location
  - Special features
  - Uses of bridge (Who is using it? What are they using it for?)

Also, individually, spend some time learning what you can about your bridge. For instance, you may consider some of the following questions:

- When was it built?
- Has your bridge been in the news?
- Are there any interesting facts about the bridge?
- Is there any fun folklore associated with the bridge?

When you return to class, your group will develop a 5-minute presentation to report your findings to the class.

Bridge Presentation Rubric Name: ________________				
Criteria	Below Standard (1 point)	Approaching Standard (2 points)	Meets or Exceeds Standard (3 points)	*Score*
HISTORICAL INFORMATION	Students provide incomplete or incorrect historical information.	Students provide some accurate historical information but did not provide enough detail on all bridges.	Students provide a detailed description of the history of their bridges. This includes age, development, historical references, and folklore.	
BASIC FACTS	Students provide little information about their bridges.	Students provide some basic facts but did not provide enough detail on their bridges.	Students give age, materials used, type, and description of overall structure of their bridges.	
FUNCTION	Students fail to describe the usage and special features of their bridges.	Students provide some description of the function of their bridges, but fail to describe how their bridges affect the communities in which they are found.	Students adequately describe the function and purpose of each bridge. This may include usage, impact on community, and special features.	
OVERALL PRESENTATION	Not all team members participate.	All team members participate in the presentation; however, presentation is disorganized and difficult to follow.	Presentation is clear, organized, and engaging. It is clear that all students participated in the development and presentation of the project.	
TOTAL SCORE: ________ COMMENTS:				

Scaling Bridges Checklist

Name: ______________________________

Items	Yes or No	Comments
STUDENT INCLUDED AN ACCURATE SCALE DIAGRAM OF THE BRIDGE.		
STUDENT INCLUDED AN ACCURATE 3-D MODEL OF THE BRIDGE.		
STUDENT DESCRIBED SIMILARITIES AND DIFFERENCES BETWEEN THE BRIDGE AND CLASSMATES' BRIDGES.		

Multimedia Rubric—Science: Mineral Resources in the State

Name: ________________________________

Criteria	Below Standard (1 point)	Approaching Standard (2 points)	Meets Standard (3 points)	Exceeds Standard (4 points)	*Score*
TOPICS COVERED	Only superficially covered any of the topics.	Sufficiently covered topics, but missed two or three of them.	Sufficiently covered most items for the types of minerals found in the state and their uses for road/bridge construction.	Thoroughly covered all items that explain the types of minerals found in the state and their uses for road/bridge construction.	
INFORMATION	Not enough information or evidence provided to be informative.	Only superficial information provided, some evidence missing.	Sufficient information and evidence provided, but some could be more thoroughly covered.	Information was thorough and well explained with provided evidence.	
PRESENTATION	Not presented in a way to form a convincing argument for audience.	Information is mostly accurate, but not well presented for audience.	Information is accurate and clearly presented for audience to understand.	Presentation is original, appealing, accurate, well thought out, and clearly presented so that audience can understand.	
SOURCES USED	Used fewer than two resources to gather information.	Used at least two reliable resources.	Used at least three reliable resources.	Used at least four reliable resources.	
SOURCE CITATIONS	There are significant errors in citations and missing information.	There is information missing, formatting errors, and some sources not cited.	All sources are cited, but there are some formatting errors.	All sources are cited correctly.	

TOTAL SCORE: _______

COMMENTS:

Lesson Plan 2: Beam Bridges

In this lesson, students begin to take a closer look at beam bridges. As students explore the basics of beam bridges, they experiment, collect data, and write functions that model the data they collect in their experiments. Finally, students use their models in a real-world setting.

ESSENTIAL QUESTIONS

- What are the limitations of beam bridges?
- What are the forces that act on a beam bridge?
- How can we use linear models to help us make predictions about what will happen?
- What roles have bridges played in history?
- What are various contexts in which the word *bridge* has been used metaphorically?
- What are differences and similarities between sedimentary, igneous, and metamorphic rocks?

ESTABLISHED GOALS AND OBJECTIVES

At the conclusion of this lesson, students will be able to do the following:

- Collect and organize data through experimentation.
- Interpret data and write a linear equation that best fits the data.
- Use a linear model to solve problems in a real-world context.
- Describe the historical impact of bridges.
- Explore metaphorical uses of the word *bridge*.
- Identify differences and similarities between sedimentary, igneous, and metamorphic rocks.

TIME REQUIRED

- 3 days (approximately 45 minutes each day; see Table 3.7, p. 39)

MATERIALS

- STEM Research Notebooks
- Chart paper
- Access to computers or other web-enabled devices
- Handouts (attached at the end of this lesson)
- 3 × 5 inch card stock (20 pieces per group)
- 2 wooden blocks per group; each block should be 2 × 4 × 6 inches (a 12 foot long 2 × 4 inch piece of lumber cut into 6 inch pieces; blocks are used to make the beams in the Beam Bridge Penny Challenge)
- Paper cups
- 100 pennies, washers, or other equivalent (per group)
- Computer and projector
- Graphing utility (e.g., graphing calculator, GeoGebra, Desmos)
- 2 half-gallon paper milk cartons per group; the tops should be cut off of the milk cartons
- Water
- Permanent marker (1 per group)
- Plaster of paris
- Water-filled balloon (1 per group)
- Ribbon measuring tapes (1 per group)
- Mass balances (1 per group, if available)
- Chisel
- Hammer
- Large nails (1 per group)
- Safety glasses or safety goggles, nonlatex aprons
- Assortment of 6–8 rock samples (the sample of rocks needs to include at least two each of sedimentary, metamorphic, and igneous rocks).

SAFETY NOTES

1. All laboratory occupants must wear safety glasses or goggles during all phases of this inquiry activity.
2. Report spills immediately and avoid walking in areas where water has been spilled on the floor.
3. Keep away from electrical sources when working with water because of the potential shock hazard.
4. Use caution when working with sharps (e.g., nails, rocks) to avoid cutting or puncturing skin.
5. Use precautions when working with hand tools (hammer): Can cause body injury.
6. Make sure all materials are put away after completing the activity.
7. Wash hands with soap and water after completing this activity.

Additional Safety Considerations

You should develop an internet/blog protocol with students if guidelines are not already in place. Since students will use the internet to acquire the needed data for their research, you should monitor students' access to the internet to ensure that students are only accessing websites that you clearly identified. Further, you should inform parents or guardians that students will create online multimedia presentations of their research and that you will closely monitor these projects. It is recommended that you not allow any website posts created by students to "go public" without first approving them.

During this lesson, students are provided with a variety of classroom materials to be used to explore the concepts in the lesson. You are encouraged to establish appropriate guidelines for distribution, use, and cleanup of the materials with your students. Inappropriate use of the classroom materials can create an unsafe environment and detract from student learning.

CONTENT STANDARDS AND KEY VOCABULARY

Table 4.4 (p. 72) lists the content standards from the *NGSS*, *CCSS*, and the Framework for 21st Century Learning that this lesson addresses, and Table 4.5 (p. 75) presents the key vocabulary. Vocabulary terms are provided for both teacher and student use. Teachers may choose to introduce some or all of the terms to students.

Table 4.4. Content Standards Addressed in STEM Road Map Module Lesson 2

NEXT GENERATION SCIENCE STANDARDS

PERFORMANCE EXPECTATION

- MS-ESS3-1. Construct a scientific explanation based on evidence for how the uneven distributions of Earth's mineral, energy, and groundwater resources are the result of past and current geoscience processes.

SCIENCE AND ENGINEERING PRACTICES

Developing and Using Models

Modeling in 6–8 builds on K–5 experiences and progresses to developing, using, and revising models to describe, test, and predict more abstract phenomena and design systems.

- Develop and use a model to describe phenomena.

Analyzing and Interpreting Data

Analyzing data in 6–8 builds on K–5 experiences and progresses to extending quantitative analysis to investigations, distinguishing between correlation and causation, and basic statistical techniques of data and error analysis.

- Analyze and interpret data to provide evidence for phenomena.

Constructing Explanations and Designing Solutions

Constructing explanations and designing solutions in 6–8 builds on K–5 experiences and progresses to include constructing explanations and designing solutions supported by multiple sources of evidence consistent with scientific ideas, principles, and theories.

- Construct a scientific explanation based on valid and reliable evidence obtained from sources (including the students' own experiments) and the assumption that theories and laws that describe the natural world operate today as they did in the past and will continue to do so in the future.

Scientific Knowledge Is Open to Revision in Light of New Evidence (Nature of Science practice)

- Science findings are frequently revised and/or reinterpreted based on new evidence.

DISCIPLINARY CORE IDEAS

ESS2.A: Earth's Materials and Systems

- All Earth processes are the result of energy flowing and matter cycling within and among the planet's systems. This energy is derived from the sun and Earth's hot interior. The energy that flows and matter that cycles produce chemical and physical changes in Earth's materials and living organisms.

Continued

Table 4.4. *(continued)*

ESS3.A: Natural Resources

- Humans depend on Earth's land, ocean, atmosphere, and biosphere for many different resources. Minerals, fresh water, and biosphere resources are limited, and many are not renewable or replaceable over human lifetimes. These resources are distributed unevenly around the planet as a result of past geologic processes.

ETS1.C: Optimizing the Design Solution

- The iterative process of testing the most promising solutions and modifying what is proposed on the basis of the test results leads to greater refinement and ultimately to an optimal solution.

CROSSCUTTING CONCEPTS

Stability and Change

- Explanations of stability and change in natural or designed systems can be constructed by examining the changes over time and processes at different scales, including the atomic scale.

Cause and Effect

- Cause and effect relationships may be used to predict phenomena in natural or designed systems.

Influence of Science, Engineering, and Technology on Society and the Natural World

- All human activity draws on natural resources and has both short- and long-term consequences, positive as well as negative, for the health of people and the natural environment.

COMMON CORE STATE STANDARDS FOR MATHEMATICS

MATHEMATICAL PRACTICES

- MP1. Make sense of problems and persevere in solving them.
- MP3. Construct viable arguments and critique the reasoning of others.
- MP4. Model with mathematics.
- MP5. Use appropriate tools strategically.
- MP6. Attend to precision.
- MP7. Look for and make use of structure.

MATHEMATICAL CONTENT

- 8.EE.B.5. Graph proportional relationships, interpreting the unit rate as the slope of the graph. Compare two different proportional relationships represented in different ways. For example, compare a distance-time graph to a distance-time equation to determine which of two moving objects has greater speed.

Continued

Table 4.4. *(continued)*

MATHEMATICAL CONTENT *(continued)*

- 8.EE.C.7.B. Solve linear equations with rational number coefficients, including equations whose solutions require expanding expressions using the distributive property and collecting like terms.
- 8.F.B.5. Describe qualitatively the functional relationship between two quantities by analyzing a graph (e.g., where the function is increasing or decreasing, linear or nonlinear). Sketch a graph that exhibits the qualitative features of a function that has been described verbally.

COMMON CORE STATE STANDARDS FOR ENGLISH LANGUAGE ARTS

READING STANDARDS

- RL.8.1. Cite the textual evidence that most strongly supports an analysis of what the text says explicitly as well as inferences drawn from the text.

WRITING STANDARDS

- W.8.1. Write arguments to support claims with clear reasons and relevant evidence.
- W.8.1.A. Introduce claim(s), acknowledge and distinguish the claim(s) from alternate or opposing claims, and organize the reasons and evidence logically.
- W.8.1.B. Support claim(s) with logical reasoning and relevant evidence, using accurate, credible sources and demonstrating an understanding of the topic or text.
- W.8.1.C. Use words, phrases, and clauses to create cohesion and clarify the relationships among claim(s), counterclaims, reasons, and evidence.

SPEAKING AND LISTENING STANDARD

- SL.8.2. Analyze the purpose of information presented in diverse media and formats (e.g., visually, quantitatively, orally) and evaluate the motives (e.g., social, commercial, political) behind its presentation.

LANGUAGE STANDARDS

- L.8.5. Demonstrate understanding of figurative language, word relationships, and nuances in word meanings.
- L.8.5.A. Interpret figures of speech (e.g., verbal irony, puns) in context.
- L.8.5.B. Use the relationship between particular words to better understand each of the words.

FRAMEWORK FOR 21ST CENTURY LEARNING

- Interdisciplinary Themes: Environmental Literacy; Civic Literacy
- Learning and Innovation Skills: Creativity and Innovation; Critical Thinking and Problem Solving; Communication and Collaboration
- Information, Media, and Technology Skills: Information Literacy; Media Literacy; ICT Literacy
- Life and Career Skills: Flexibility and Adaptability; Initiative and Self-Direction; Social and Cross-Cultural Skills; Productivity and Accountability; Leadership and Responsibility

Table 4.5. Key Vocabulary for Lesson 2

Key Vocabulary	Definition
agent of erosion	something moving, such as water, ice, or wind, that can transport sediment
beam	"a rigid, usually horizontal structural element" (PBS 2001)
beam bridge	"a simple type of bridge, composed of horizontal beams supported by vertical posts" (PBS 2001)
cement matrix	a combination of very fine particles that support the larger grains in sandstone; forms when the rock is solidifying and acts like a glue that holds the individual particles of the sandstone together
cementation	the process by which minerals dissolve in water cement sediment together during the formation of sedimentary rock
chemical weathering	the breaking down of minerals or rocks due to interactions with water or some other solution
compaction	the process by which layers of sediment are squeezed together during the formation of sedimentary rock
compression	"a pressing force that squeezes material together" (PBS 2001)
deck	"supported roadway on a bridge" (PBS 2001)
deposition	process of depositing sediments in a new location; sediments can be moved by wind, ice, water, or gravity
erosion	the process by which rock particles are worn away by wind, water, ice, or gravity and move to another location
force	"any action that tends to maintain or alter the position of a structure" (PBS 2001)
ice wedging (or frost action)	the mechanical weathering of rocks that results from the freezing of water
igneous rock	a type of rock that forms from the solidification of molten rock material (i.e., lava or magma)
mechanical weathering	the breaking of rocks into smaller pieces
metamorphic rock	a type of rock that has been changed by pressure, heat, and chemical processes; these changes typically take place when the rock is buried deep below the Earth's surface
obsidian	volcanic rock formed by rapidly solidifying lava; dark, hard, and glasslike rock

Continued

Table 4.5. *(continued)*

Key Vocabulary	Definition
pier	"a vertical supporting structure, such as a pillar" (PBS 2001)
pumice	volcanic rock formed when gas-rich lava solidifies rapidly; light and porous rock
sag	amount that a bridge dips below the horizontal
sedimentary rock	a type of rock that forms when small particles of sediment (e.g., mud, sand, pebbles) accumulate on land or in water
span	"the distance the bridge extends between two supports" (PBS 2001)
tension	a stretching force that pulls materials apart
vesicle	cavity in volcanic rock formed by the presence of gas bubbles in the lava as it solidifies

TEACHER BACKGROUND INFORMATION

Mathematics

One of the earliest forms of bridge construction is a beam bridge. In its most basic form, a beam bridge is a horizontal beam resting on two supports or piers. Beam bridges are simple, relatively strong across short span lengths, and inexpensive compared with other bridge designs. As a result, beam bridges are the most common design for bridges. Both compression and tension forces are present in beam bridges. As the beam presses downward on the piers, compression forces cause the top of the beam to squeeze inward while tension forces cause the bottom of the beam to stretch outward. As the span length between supports increases, compression and tension forces increase, causing the beam to weaken. In the span length activity, the weakening of the beam can be seen by the amount of *sag* that is present in a piece of cardstock. It is important to inform students that while real building materials will sag as compression and tension forces increase, it is exaggerated in this experiment. Using cardstock allows students to see the weakening of the beam as they add weight. Because the beam weakens as span length increases, beam bridges are usually not used for spans longer than 250 feet. For more information about beam bridges, visit the following websites:

- *www.brighthubengineering.com/structural-engineering/46079-beam-bridges-history-construction-and-future*
- *www.brighthubengineering.com/structural-engineering/65074-characteristics-of-beam-bridges*

Science

Rocks and minerals can be differentiated by the number of constituent substances. Specifically, minerals are composed of only a single substance while rocks can be composed of one or more minerals. If a rock is composed of many minerals, then those minerals are held together with a cement-like substance called a *cement matrix.*

Types of Rocks. There are three basic types of rocks: (1) *sedimentary,* (2) *igneous,* and (3) *metamorphic.* Sedimentary rocks form when particles of sediment (which can include bits of sand, pebbles, shell, and other fragments) accumulate in layers over long periods of time and harden into rock. Sandstone, limestone, and conglomerates are examples of sedimentary rocks. Igneous rocks form when magma cools and hardens. The rate at which the magma cools determines the type of igneous rock that forms. Sometimes magma cools slowly inside the Earth and other times it will erupt onto the surface from volcanoes (when magma is on the Earth's surface, it is called *lava*). When lava cools quickly, no opportunity arises for crystals to form and the resulting rock, *obsidian,* is glasslike and shiny. Sometimes gas bubbles are trapped in the rock while it cools and form tiny holes and spaces in the rock called *vesicles; pumice* is an example of an igneous rock with vesicles. Metamorphic rocks result from preexisting sedimentary or igneous rocks that are buried under the surface of the Earth and subject to intense heat and pressure. Gneiss and marble are examples of metamorphic rocks. For more information on rocks and minerals, please see *www.geology.com.*

Weathering, Erosion, and Deposition. Weathering is a process by which large pieces of rocks become smaller pieces by being dissolved, worn away, or broken into smaller pieces (or grains or clasts). Weathering processes are classified as mechanical or chemical. *Mechanical weathering,* also called *physical weathering,* is the physical breakdown of rocks into smaller pieces by forces that are exerted on the rocks. *Frost action* (or *ice wedging*) is one type of mechanical weathering that occurs due to the expansion of water upon freezing since water has a 9% volumetric expansion upon freezing. *Chemical weathering* results when minerals interact with a solution (typically water) and there is a change in the stability of the minerals present in the rock.

Weathering and erosion are NOT synonymous. Weathering involves the breaking down of the rock. As soon as that rock particle moves, it is called *erosion.* For more information on weathering, please see the following U.S. Geological Survey websites:

- *http://education.usgs.gov/lessons/schoolyard/RockDescription.html*
- *http://geomaps.wr.usgs.gov/parks/misc/gweaero.html*

ELA

Students will explore bridge metaphors and bridge "stories" through nonfiction texts in this lesson. A brief overview of metaphors and nonfiction text structures and features are therefore provided here.

Metaphors. A metaphor maps the qualities of one object or concept onto another to offer insight into the qualities of the new object. Bridge metaphors are pervasive in our language because a bridge stands as a symbol of powerful human experiences. For example, a bridge can represent an in-between place, a place of decision, a place of opportunity, what lies ahead, mourning, and what is left behind. It could also be a precarious place. It can be a physical symbol of human toil and ingenuity, of overcoming extensive barriers. Bridges may also be places of power. They affect commerce, grow communities, connect cultures, and create opportunities. The absence of a bridge, of course, can result in just the opposite, including isolation and diminished resources. Bridges may be sources of power; if you can control access to a bridge (e.g., through tolls or physical blockades), you can control access to resources and opportunities.

As students explore metaphorical uses of bridges, they will encounter phrases that are used or labeled as *idioms* and may question that difference. An idiom is habitual speech that may not use the literal meanings of the words. Idioms are almost always grounded in metaphor, and that relationship can be "teased out" by unpacking the idiom. For example, a common idiom is "you are driving me up the wall." To unpack that idiom metaphorically, consider the image of someone being chased and cornered: There is no place else to go than up the wall.

Nonfiction Text Structures and Features. Nonfiction writers employ a variety of text structures and features to organize their writing and effectively convey meaning. Readers who are aware of these structures are able to quickly see past the surface details to understand at a deeper, more critical level.

Text structure refers to the way ideas are laid out and related. Some common text structures (Sanders and Moudy 2008; Harvey and Goudvis 2007) include the following:

- *Cause and effect:* The author highlights how one action or event directly causes another.
- *Problem and solution:* The author calls attention to a problem, but goes on to show the process of coming to a solution.
- *Question and answer:* The use of this structure highlights inquiry, a vital process in reading, writing, and learning in general. In the text, a question is posed for the reader to ponder, then the (or an) answer is addressed.

- *Chronological/sequential:* The author explains an event or process with steps in order.
- *Compare and contrast:* Comparing and contrasting texts help us understand one thing by showing how it is like or different from another (perhaps more familiar) thing.
- *Descriptive:* Authors use rich detail and sensory language to recreate a place or event so that the reader comes to understand what it is like.
- *Narrative informational:* Often employed in nonfiction texts for the very young or in biographies and autobiographies, texts employing this structure read like stories.

An author may use one structure primarily throughout the piece, or may employ several in a more complex text. When we teach learners to recognize these structures in a text, we are helping them more easily comprehend the material. For instance, recognizing a problem and solution structure in a text allows the reader to attend to the scientific process and the outcome, as well as to home in on the concept that scientists identify problems and consider possible solutions until the problem is addressed. Basically, we are helping learners see past the details to really get at the important information.

Text features function much like road signs. They help readers navigate the text and call attention to important features that enrich the experience. Rich with photos, graphics, and sidebars, many nonfiction texts are not meant to be read in a consistent left-to-right, line-by-line manner. Instead, a reader may begin in different places on a page looking, for instance, at a captioned photo or a diagram first before reading the more traditional text. Or, while readers usually read a fiction book from front to back, they are more likely to sample sections of a nonfiction book, searching for particular information. Teaching readers the functions of and relationships between text features helps them get the most out of a text and, when they are writing, to produce complex and meaningful work. Common nonfiction text features (adapted from Harvey and Goudvis 2007) include the following:

- Organizational supports such as a table of contents or an index that help readers find information
- Titles and headings that alert the reader to what a section of text is about and support skimming and scanning for information
- Varied fonts and text effects that highlight terms or other important information for the reader
- Photographs and illustrations that clarify and augment the text

- Graphics such as charts and tables that pack a lot of information into a small space
- Glossary or other "back matter" that explains specialized vocabulary or provides context for the content
- "Textual cues"—*for example, for instance, in fact, in conclusion, most important, but, therefore, on the other hand* (Harvey and Goudvis 2007)—that call the reader's attention to important information and give clues about the text structure

COMMON MISCONCEPTIONS

Students will have various types of prior knowledge about the concepts introduced in this lesson. Table 4.6 outlines some common misconceptions students may have concerning these concepts. Because of the breadth of students' experiences, it is not possible to anticipate every misconception that students may bring as they approach this lesson. Incorrect or inaccurate prior understanding of concepts can influence student learning in the future, however, so it is important to be alert to misconceptions such as those presented in the table.

Table 4.6. Common Misconceptions About the Concepts in Lesson 2

Topic	Student Misconception	Explanation
Variables	There is confusion about which are independent and dependent variables.	An independent variable is a variable that you can control (span length of a bridge). A dependent variable is a variable that you observe and measure (weight the span will hold).
Models	Models are only in 3-D.	Models in fact can be 2-D representations such as pictures, graphs, written descriptions, blueprints, equations, and other representations, as well as a 3-D model.
	A model's usefulness is based solely on the model's physical resemblance to the object being modeled.	A physical model is a smaller or larger physical copy of an object such as a bridge. The model represents a similar object in the sense that scale is an important characteristic of the model.
Force (a push or pull between objects)	Nonmoving objects do not exert a force (e.g., a tabletop or roadway).	Stationary objects can exert forces on other objects. For example, when you roll a toy car over a tabletop, there is a frictional force between the table and the car wheels.
	If a moving object is slowing, the force that was propelling it forward is decreasing.	When a force acts on a moving object in a direction opposite the direction of motion, the moving object will slow, even if the force that was propelling the object forward continues.
Weathering, erosion, and deposition (the breaking down of a material, the movement of the material to another area)	Water and wind cannot wear away rock and other building materials.	Water and wind can wear away rock and other building materials.
	Water freezing in cracks cannot break apart rock.	Water freezing in cracks can break apart rock.
Earth materials (types and uses of rocks)	Rocks have little practical use.	Rocks are used for a wide variety of consumer and industrial purposes. For example, rocks are used in paper production, cement, household cleaners, jewelry, pencils, chalk, glass, and building materials such as bricks and kitchen countertops.
	All rocks are the same.	There are three major types of rocks (sedimentary, igneous, and metamorphic), which have a variety of mineral compositions.

PREPARATION FOR LESSON 2

Gather materials for the investigation and make copies of the student handouts attached at the end of this lesson. Test the river template on the first page of the Beam Bridge Penny Challenge handout (p. 93); support boxes may need to be adjusted to fit the size of card-stock that you use. Have a computer and projector on hand and be prepared for students to use a graphing utility, either by ensuring that they have access to graphing calculators or to graphing software or an online graphing program such as GeoGebra or Desmos.

For ELA, assemble a collection of literature about bridges for students to use to identify bridge metaphors. This may include poetry, song lyrics, short stories, or novel excerpts.

LEARNING COMPONENTS

Introductory Activity/Engagement

Connection to the Challenge: Begin each day of this lesson by directing students' attention to the driving question for the module and challenge: How can we develop a decision model to help us make a recommendation to the local department of transportation on the type of bridge to build for a given location? Ask students why bridges are so important and what impact a bridge might have on a community. Hold a brief student discussion of how their learning in the previous days' lesson(s) contributed to their ability to create their plan and build their prototype. You may wish to hold a class discussion, creating a class list of key ideas on chart paper, or you may wish to have students create a notebook entry with this information.

Driving Question for Lesson 2: How do beam bridges support the weight of trucks and cars for many, many years?

Mathematics Class: Place students into small groups and provide them with a piece of chart paper. Groups should brainstorm what they know about beam bridges and what they still need to know. Once students have finished, post the charts on the wall for all students to see. Allow students to share their ideas with the class.

STEM Research Notebook Prompt

Have students respond to the following prompts in their STEM Research Notebooks:

- *Of the bridges we have looked at so far, what types of bridges do you think were built first?*
- *Which types of bridges do you think are the easiest to build?*

Science Connection: Arrange students in small groups and have an assortment of rocks (two or three of different types) at the front of the room. Inform students that the cool thing about rocks is that they tell a story: By being a good detective and observer, the rock will give you clues as to where it has been, how it was made, and how long it has been there.

STEM Research Notebook Prompt

Have the students respond to the prompt *What is a rock?* in their STEM Research Notebooks. After students have recorded in their STEM Research Notebooks, have them share their thoughts in a small group and then with the whole class. If students haven't arrived at a concise definition of rocks, prompt them by asking the following questions:

- *What are rocks made of?*
- *Are all rocks the same?*
- *What is the relationship between rocks and minerals? Are they the same thing?*

Have students (a) compare and contrast rocks and minerals and (b) brainstorm why it is important for engineers who are going to build a bridge to understand about rocks and minerals.

STEM Research Notebook Prompt

Provide each small group of students with an assortment of six to eight rock samples (the sample of rocks needs to include at least two each of sedimentary, metamorphic, and igneous rocks). Allow students to make an observation table in their STEM Research Notebooks and then record observations of each rock. For each rock, the student needs to (a) make a sketch (with labels for clarification) and (b) record at least three observations. (See Rock Observation Rubric on p. 106 for grading criteria.) As students fill out their tables, encourage students to use what they learned about mineral identification from the previous lesson to try to identify the minerals in the rock samples.

Next, have students do an open sort of the rocks by organizing the rocks in groupings of their choice. For each rock grouping, the students list two or three common characteristics, develop a name for the rock group, and provide an explanation for their groupings. Students record this in their STEM Research Notebooks. After student groups have established their rock groups, each group elects one student to be their spokesperson. Each spokesperson will explain how the group organized its rocks. (*Note:* At this point, the students may have any number of groupings for their rocks based on the characteristics they identified.) Finally, ask all students to share their data and inform them that geologists classify rocks into three types (sedimentary, metamorphic, and igneous). Challenge the students to reorganize their rocks into three groups.

ELA Connection: Provide examples of several metaphorical or idiomatic uses of the word *bridge,* such as "I'll cross that bridge when I come to it" or "Don't burn your bridges," and talk together about what those phrases mean. For instance, ask students if we are talking about actual bridges. Review the concept of metaphor and briefly discuss the important role metaphors play in our understanding of the world and language. Point out that *bridge* is a particularly common metaphor in our language as bridges symbolize, or stand for, significant human experiences. Tell students that they are going to conduct an exploration and analysis of how we use bridges as a symbol to express important human experiences.

STEM Research Notebook Prompt

Although we know that metaphorical or idiomatic use of the word *bridges* does not directly play into a decision model for bridges, have students reflect in their STEM Research Notebooks about why thinking about these uses are important from a community perspective and why these uses might play a role in the decisions surrounding decisions to build or not build a bridge.

Social Studies Connection: Begin to research important bridges in U.S. history and roles that they have played. Assign teams of students an important bridge in the United States (e.g., the bridge out of New Orleans during Hurricane Katrina, the Brooklyn Bridge after 9-11, bridges flattened in the Northridge earthquake). Have students research the bridge and prepare a 5-minute multimedia presentation to share their findings with the class. Students should focus on the historical significance of the bridge, as well as the construction process and the type of bridge it represents (i.e., beam bridge, suspension bridge, arch bridge, or other type of bridge).

Activity/Exploration

Mathematics Class: Divide students into their project groups and distribute the Beam Bridge Penny Challenge handout. Each group will use the handout and materials provided to complete the following challenge: *On the piece of paper in front you there is a picture of small stream. Your task is to determine how many layers of cardstock would be required to build a bridge strong enough to support 100 pennies. However, due to the cost of this cardstock you will only be given 10 pieces.*

Note: **It is important to ensure the setup is such that the cardstock only overlaps the supports by approximately ¼ of an inch on each side. Too much overlap will cause the bridge not to fail.**

As students conduct their experiment, encourage them to represent their data in a meaningful way. A graphical representation of this data will help them see the linear nature of the data. Once this pattern is recognized, students can make predictions about how many layers of cardstock will be needed to hold 100 pennies.

STEM Research Notebook Prompt

Once students have completed the Beam Bridge Penny Challenge, have them answer the following prompts individually in their STEM Research Notebooks and then facilitate a class discussion:

- *What are some observations you made about the beam bridge as you completed the investigation?*
- *Do you believe this is an effective bridge design? Why or why not? What did you notice about how the bridge collapsed?*

Science Connection: Have the students share with the class how they classified the rocks into three groups. Explain to the students that there are three basic types of rocks: sedimentary, metamorphic, and igneous. Provide students with definitions of each rock type. Allow students to re-sort their rocks one final time to match the definitions of the three rock types. Check for student understanding by making sure that all rocks are properly sorted into the three types of rocks. Have students record a definition for sedimentary, ingenious, and metamorphic rocks in their STEM Research Notebooks.

STEM Research Notebook Prompt

Pose the following question: *When building a bridge, what are some uses and considerations of rocks?* Have students do a quick write in their STEM Research Notebook. As students share their answers with the class, all students should be encouraged to add to their responses.

Ask the class the following: *How do big rocks turn into smaller rocks in nature?* Give students an opportunity to discuss in smaller groups and then conduct a class discussion and record the ideas for the class. Explain to the class that roads are made of Earth materials and to better understand how roads will perform (and last), we need to have an understanding of how Earth materials perform. Also, when people build a bridge, they must consider the local geology as the bridge will be supported by the rock types found at the building site. Construction companies often employ a geological engineer to help them make decisions about how well suited the rocks are at a building site. The geological engineer will study the rocks at the site to help the company make construction decisions.

ELA Connection: How do we use the word *bridge* as a metaphor? Have students search for and collect as many metaphorical uses of the word *bridge* as possible. They might, for instance, "crowd source" the question, using face-to-face and social media conversations to brainstorm uses; explore song lyrics and poetry; and conduct internet searches. Collect their examples in class. Consider tracking frequency of response as well as the

scope of responses, that is, which examples were most easily and often discovered? Once examples are collected, discuss the meanings offered. Some possible analytical engagements might be comparing and sorting into like-meaning categories, employing "sketch to stretch" (Short and Harste 1996) to explore the relationship between actual and metaphorical meaning, and using found examples as mentor sentences to support their own written examples.

Build on student findings to collectively construct a statement about why the bridge is such a pervasive metaphor in our language. (See the Teacher Background Information section on p. 76 for guidance.) Have students use their new insights into bridge as a metaphor to explore literature (e.g., short stories, poems, song lyrics).

Social Studies Connection: Have students work in their teams to brainstorm questions for which they should find answers and address in their presentation.

Explanation

Mathematics Class: Give each student a Beam Bridge Basics handout (p. 98). Many students will have come across the bridge terms on the handout while conducting their research at the beginning of the unit. As a class, discuss these terms and develop their definitions. Encourage students to relate these terms to the Beam Bridge Penny Challenge. As students share their ideas, discuss the following questions to help connect the activity to the mathematics content:

- How did you display your data?
- What were the variables involved in this problem?
- What was the relationship between those variables?
- How did you come up with a model (equation) that related those variables?
- Does your solution make sense in the context of this problem? Why or why not?
- What factors must be considered about the beam?

Once the students have shared their findings, present them with the following problem: *The engineers have further surveyed the site and have determined that the supports are only able to support the weight that is equivalent to four pieces of cardstock! To make matters worse, the soil in the river is too unstable to allow for the construction of an additional support. However, the bridge they construct must still be able to support* at least *the same amount of weight!*

Give each group eight new pieces of cardstock. Using the same setup, students must find a way to construct a bridge out of these materials that can support 100 pennies. From the earlier experiment, students should recognize that four pieces of cardstock layered on top of each other are not strong enough to support 100 pennies. After a few minutes, give each group a small piece of corrugated cardboard. This may not be used as building materials but is meant only for inspection. Ask students the following questions as each group designs a solution to the task:

- What is cardboard made out of?
- If it is only made of paper, why is it so much stronger?
- What is it about its design or structure that makes it so strong?
- How could we use these ideas to improve our beam design?

At this point, students should see that there is paper folded in between two layers of paper. *(Note: Here is a possible solution to creating a stronger bridge deck using only 4 pieces of 3 × 5 inch cardstock: Take two pieces of cardstock and fold them like an accordion lengthwise and put them together in between the remaining two notecards. The result should look like corrugated cardboard.)*

Once students have constructed their new style of beam bridge, have them determine whether it is able to support the 100 pennies and give them the Beam Bridge—New and Improved handout (p. 97).

STEM Research Notebook Prompt

In their STEM Research Notebooks, have students draw and describe their solution to the task and then reflect on the following prompts:

- *What have you learned about beam bridges during these investigations?*
- *Why is it important for engineers to continue to explore new design solutions?*

Science Connection: Inform students that they will investigate one type of weathering that has a major impact on roads and bridges. The students will investigate the effect of ice wedging on roads (and Earth materials) using the Ice Wedging Activity handout (p. 103). In this activity, students use plaster of paris and a water-filled balloon to observe the effects of the freezing and thawing of water on "rock" (the plaster of paris). By comparing a "rock" with the water-filled balloon inside it with a "rock" with no water balloon, students should observe that the water balloon expanded when it froze, causing cracks (ice wedging) in their "rocks."

Introduce the idea that ice wedging is an example of physical weathering. Discuss with the class how ice wedging might affect roadways and sidewalks and what measures prevent ice wedging (e.g., sealing cracks in pavements, using building materials that have the capacity to expand).

STEM Research Notebook Prompt

Ask the students to reflect on what other types of weathering they may know about. Once the students have had a few minutes to record their thoughts in their STEM Research Notebooks, go over other types of weathering and erosion with the class (this should be a review from previous grades).

ELA Connection: Have students share insights they have gained from their research into the life stories of particular bridges. Explain that they will be using that research as a base for creating and publishing a multimedia text about the life of their chosen bridge. Provide a variety of examples of place "life stories" by employing varied media. Examples of possible book resources are included in the Additional Resources section at the end of this chapter. A wide variety of video documentary resources can be found on the History Channel website at *www.history.com.*

Allow students to explore the resources and engage them in dialogue about possible approaches and formats for sharing the biographies of their bridges. Other forms to consider with students include graphic novels and other forms of visual artwork, including architectural modeling, so provide examples of those forms as well. Consider also providing examples of storytelling that mixes forms of writing and art such as *The Mangrove Tree: Planting Trees to Feed Families* (Roth and Trumbore 2011). Allow students to explore the resources and engage them in dialogue about possible approaches and formats for sharing the biographies of their bridges.

Social Studies Connection: As students complete their presentations, allow them to share with their classmates. Students should record their thoughts in their STEM Research Notebooks as other groups present.

Elaboration/Application of Knowledge

Mathematics Class: Present the following problem: *Using our new construction techniques, engineers from the local department of transportation were able to decide on a design for their new bridge. In fact, due to the results that we produced in our last investigation, they have decided to enlist our help in solving another problem. They want to know the effect that span length has on a beam bridge. There are a number of other sites that they would like to construct bridges on, but wonder if an increased distance will have an impact.*

Give each group a Beam Bridges—Effect of Span Length handout (p. 100). The goal for this investigation is for students to develop their own method of answering this question. The ideal scenario is for students to look at span length vs. sag. While the groups are working, find a group that used these variables and record their data in graphing software that can be displayed for the class (possible graphing programs include graphing calculators, GeoGebra [*www.geogebra.org/download*], or Desmos [*www.desmos.com/calculator*]). As the groups finish their experiments, allow the groups to share their findings. If possible, have a group share that looked at span length versus sag. Discuss the trends the group observed. Consider the following questions to facilitate discussion:

- As span length increases, what happens to the amount of sag?
- Is this relationship linear?
- Could we determine a model that relates these two variables? Why or why not?
- What causes a bridge to sag? Think about the forces that are acting on the bridge. What is the main source of weight as the beam gets longer?
- Does this mean that beam bridges cannot be used to span long distances?

Show students a picture of the bridge that spans Lake Pontchartrain in Louisiana. Tell the class it is the longest bridge in the United States and spans more than 24 miles, and then ask the following questions:

- What type of bridge does this appear to be?
- What did engineers have to do to make a beam bridge span this distance?

Next, give each group the Beam Bridges—Span Length and Number of Support Columns handout (p. 102) and present the following challenge to the class: *The department of transportation was interested in the results we found for span length and sag. The engineers had looked into this issue previously and determined that the maximum span length of any one section should not exceed 250 feet. Based on this information, the department of transportation would like you to write an equation that gives the number of support columns that will be needed so that the engineers can construct a beam bridge across any distance.*

STEM Research Notebook Prompt

Ask students to answer the following questions in their Research Notebooks: *Beam bridges are cheap and now we have shown that they can be built to span any length. Does this mean that we should always use a beam bridge? What are some of its limitations?*

Science Connection: Tell students that they will now study images from roads and bridges and apply what they know about weathering and erosion to roadways. Show images of road weathering to class. Pictures of weathering on local roads and bridges can be used or you can use images attached at the end of this lesson. Examples include the following:

- Pothole
- Road erosion

- Cracked asphalt
- Root wedging

STEM Research Notebook Prompt

Ask the students to write down what they observe in the images and what they infer as the cause of what they observe in the images.

Have students share their ideas in their groups and then have them share with the whole class. Provide groups of students with chart paper. Groups should split their chart into two columns. In the first column, they should record implications of weathering in the design and function of a bridge. The second column should be questions they need to still find out about the implications of weathering that might play a role in their decision model. Post these charts in the room and strive to come back to these throughout the unit. Students should also record their group's chart and ideas from others in their STEM Research Notebooks.

ELA Connection: Review nonfiction writing structures and features and engage composition groups in discussions of what structures and features might best serve them in composing their biographies. Have students work in authoring teams *with teacher guidance* to write a proposal for the format of their published bridge biographies. See the Bridge Biography Composition and Production Proposal handout at the end of Lesson 3 (p. 140). The proposal should include a story board showing order and significant events to be included, a description of the ways students will employ various media in their storytelling, a description of author roles (who will be doing what), and a plan for how they will accomplish the task (including a schedule). Provide students with a schedule of class sessions that will be provided for work so they can use that in their planning.

Social Studies Connection: Once all students have had the opportunity to present, allow them to make adjustments to their presentations based on feedback and share them on the class website. Students could also interview a parent or other adult about important bridge moments in their memory. They could write a living history account of their interviewee's memories and experiences.

Finally, reorganize teams. Form new teams with members that each worked on a different bridge, and have each group develop a two-column chart and list the similarities and differences of the various bridges, including design, purpose, and impact.

Evaluation/Assessment

Students may be assessed on the following performance tasks and other measures listed

Performance Tasks

- End of Lesson Assessment: Have students create a cluster web on beam bridges. They should write *Beam Bridge* in the center circle and then add details about beam bridges they have learned in the surrounding circles. Students can add as many circles as they need to the cluster web. A sample template can be found online at *www.educationoasis.com/printables/graphic-organizers/cluster-web-1*.
- Social Studies Bridge Multimedia Presentation Rubric
- Rock Observation Rubric

Other Measures

- STEM Research Notebook entries. You should regularly read and respond to students in their STEM Research Notebooks. Your response should not indicate whether students' entries are right or wrong. Instead, include comments or questions that will push and stretch students' thinking and can aid students in moving toward development of their decision models.
- Learning activity responses: Bridge Beam Penny Challenge, Beam Bridges—Effect of Span Length, and Other Beam Bridge Facts handouts.

INTERNET RESOURCES

Graphing calculators

- *www.desmos.com/calculator*
- *www.geogebra.org/download*

Information about beam bridges

- *www.brighthubengineering.com/structural-engineering/46079-beam-bridges-history-construction-and-future*
- *www.brighthubengineering.com/structural-engineering/65074-characteristics-of-beam-bridges*

Information about rocks and minerals

- *www.geology.com*

Information about weathering

- *http://education.usgs.gov/lessons/schoolyard/RockDescription.html*
- *http://geomaps.wr.usgs.gov/parks/misc/gweaero.html*

Video documentary resources for place "life stories"

- *www.history.com*

Sample cluster web template

- *www.educationoasis.com/printables/graphic-organizers/cluster-web-1*

Name: ______________________________

STUDENT HANDOUT, PAGE 1

BEAM BRIDGE PENNY CHALLENGE

TEMPLATE

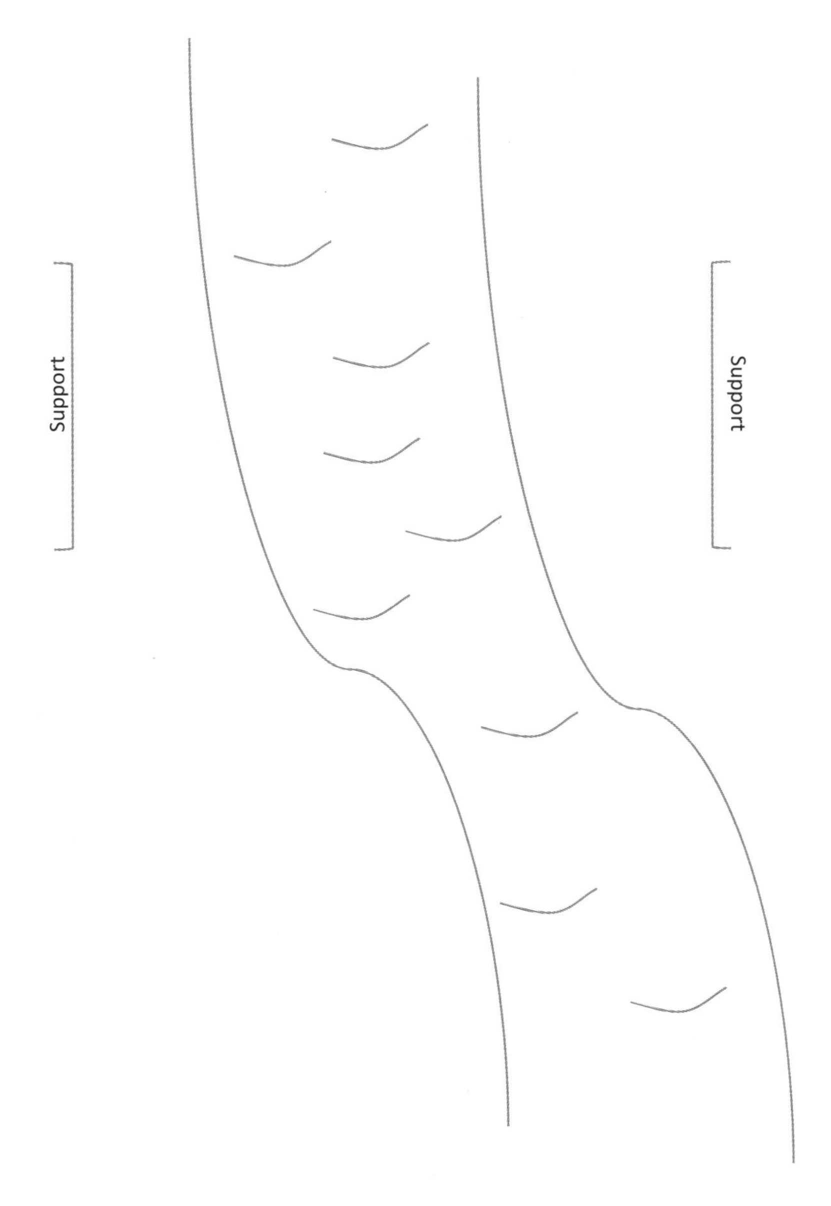

Note: It is important to ensure the setup is such that the cardstock only overlaps the supports by approximately ¼ of an inch on each side. Too much overlap will cause the bridge not to fail.

Name: ______________________________

STUDENT HANDOUT, PAGE 2

BEAM BRIDGE PENNY CHALLENGE

PROBLEM

The local department of transportation wants to construct a beam bridge across a local river. The bridge must be capable of handling heavy traffic and is located on a stretch of road that consistently experiences traffic jams. Because of this, the bridge must be very strong and be able to support a very large unmoving weight. Your task is to help them determine the strength of their materials and how much of it will be needed to build their bridge.

MATERIALS

- Cardstock: The cardstock you have been provided has been specially formulated to simulate layers of building material. One layer of cardstock has the same load capacity as one layer of building material. Unfortunately, due to the extreme high cost of developing this special cardstock, you will only be given 10 pieces. Please use them wisely!
- Pennies: The city's civil engineers have concluded that if the bridge you create can support 100 pennies, it will be strong enough to support the necessary amount of traffic when the real bridge is built.
- Wooden blocks
- Paper cup
- Safety glasses or safety goggles

SAFETY NOTES

1. All laboratory occupants must wear safety glasses or goggles during all phases of this inquiry activity.
2. Make sure all materials are put away after completing the activity.
3. Wash hands with soap and water after completing this activity.

PROCEDURE

1. Use the template on the first page of the Beam Bridge Penny Challenge to set up the test site. Be sure to place the objects in the labeled positions. This template has been created to perfectly match the site.
2. Place a paper cup in the middle of your bridge. This will allow you focus where you want to place your weight without worrying about the pennies falling off.
3. Test your bridge at increasing thicknesses. Make sure to add one penny at a time. As you are adding weight, be sure to observe how the bridge behaves. How does it fail? Are there weak points? Is it a gradual fail?
4. Record and interpret data.

Name: ______________________________

STUDENT HANDOUT, PAGE 3

BEAM BRIDGE PENNY CHALLENGE

Layers of Cardstock	Number of Pennies *Before* Collapse	Observations

INVESTIGATION QUESTIONS

1. What are the variables involved in this problem?

2. How can you organize the data to give a clearer picture of what is happening?

Name: ____

STUDENT HANDOUT, PAGE 4

BEAM BRIDGE PENNY CHALLENGE

3. Describe the relationship between your variables.

4. How could you construct a model (equation) that relates the variables?

5. With your group, determine how many layers of building material (cardstock) would be needed to support the required amount of weight.

6. Do you believe this is an acceptable number? Why or why not?

Name: ______________________________________

STUDENT HANDOUT

BEAM BRIDGE—NEW AND IMPROVED

Answer the following questions:

1. What is it about this structure that makes it stronger?

2. Why do you think this happens?

3. How much weight will your new style of bridge support?

4. As a percentage, how much stronger is this beam bridge than one with four pieces of cardstock layered on top of one another?

5. If we were to create an equation that modeled how much weight could be supported as we added cardstock, how would it compare with our original line of best fit? How would it be the same? How would it be different?

Name: ______________________________

STUDENT HANDOUT

BEAM BRIDGE BASICS

Write the definition of each of the following terms. Be sure to label the picture below.

Span:

Beam:

Deck:

Support (Pier):

Write the definition of the following forces. Be sure to label these forces on the diagram below.

Compression:

Tension:

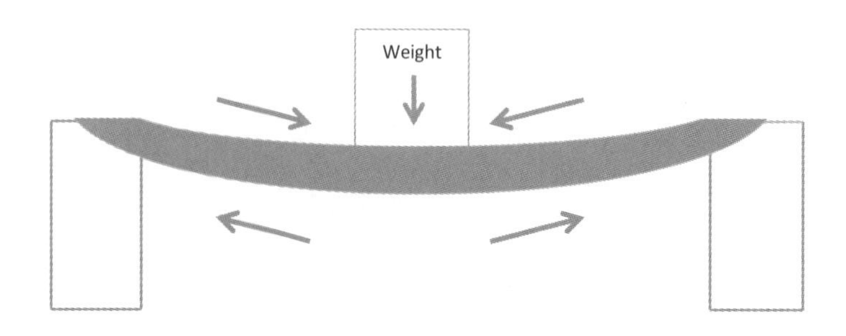

Name: ____

STUDENT HANDOUT

OTHER BEAM BRIDGE FACTS

Answer the following questions:

1. What materials in your investigation represented the beam, deck, and support?

2. Did you see the forces we discussed as you worked through your investigation?

3. What effects did the different forces have?

4. What properties must building materials have to account for the forces that act on a bridge?

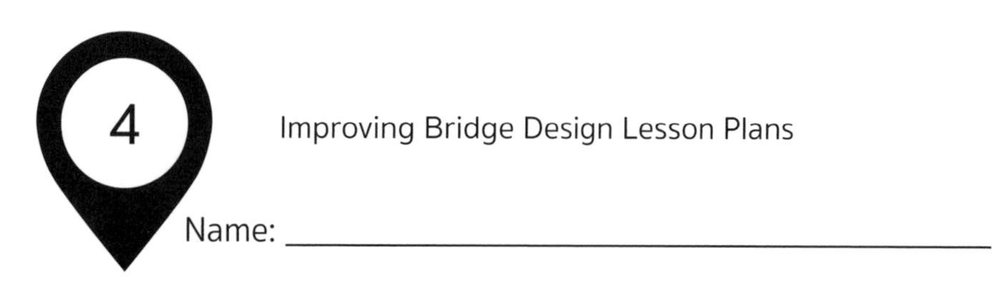

Name: ____

STUDENT HANDOUT, PAGE 1

BEAM BRIDGES—EFFECT OF SPAN LENGTH

PROBLEM

Because of your work on solving the weight issues for the local department of transportation, the department has decided to enlist you to help solve a new problem. There are a number of other sites on which the department would like to construct a beam bridge but there are concerns about the distance. The engineers would like you to determine the effect that span length has on a beam bridge so they can decide whether a beam bridge is suitable for these sites.

MATERIALS

- Wooden blocks
- Cardstock (This cardstock has been specially formulated to behave like the building materials that will be used. It's expensive so please use wisely.)
- Ruler or measuring tape
- Safety glasses or safety goggles

SAFETY NOTES

1. All laboratory occupants must wear safety glasses or goggles during all phases of this inquiry activity.
2. Make sure all materials are put away after completing the activity.
3. Wash hands with soap and water after completing this activity.

PROCEDURE

1. Identify the variables involved in this problem.
2. With your group, develop a method for determining what effect span length has on a beam bridge. (As you experiment, remember to keep as many variables as consistent as possible.)
3. Collect and record your data.
4. Using a graph, display your data and write a conclusion based on what you notice.

DATA TABLE

(Be sure to fill in the variables in the first row.)

Name: ______________________________

STUDENT HANDOUT, PAGE 2

BEAM BRIDGES—EFFECT OF SPAN LENGTH

INVESTIGATION QUESTIONS

1. What two variables are you comparing to solve this problem?

2. How are these variables related? Describe the relationship between these two variables using complete sentences. Why do you think this relationship occurs?

3. Is the relationship between these two variables linear? What does that tell us?

4. Could you write a model (equation) that relates these two variables? If so, write it here.

5. What conclusion can you make about beam bridges and span lengths? What does this mean for the department of transportation?

Name: ____________________________

STUDENT HANDOUT

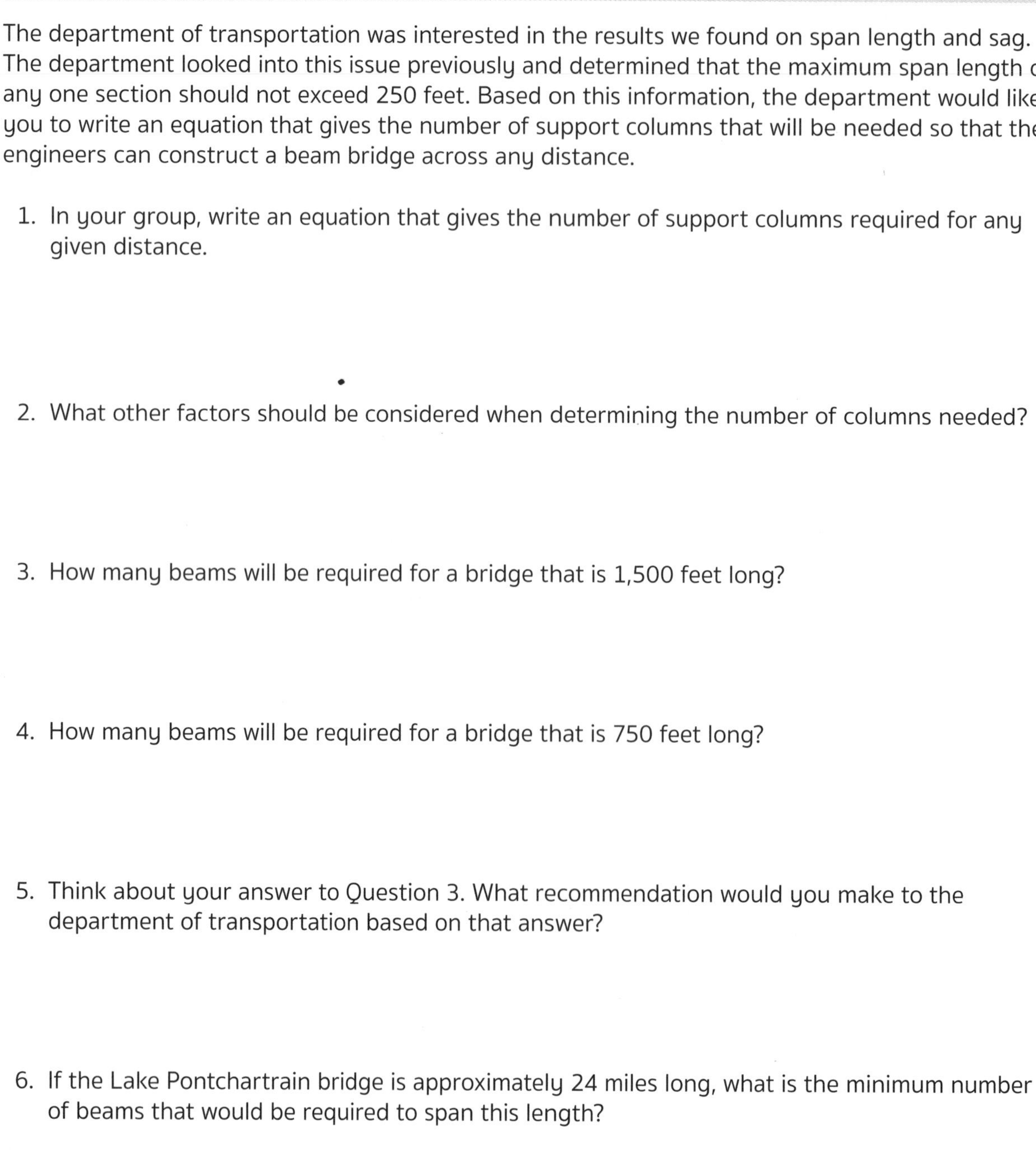

BEAM BRIDGES—SPAN LENGTH AND NUMBER OF SUPPORT COLUMNS

The department of transportation was interested in the results we found on span length and sag. The department looked into this issue previously and determined that the maximum span length of any one section should not exceed 250 feet. Based on this information, the department would like you to write an equation that gives the number of support columns that will be needed so that the engineers can construct a beam bridge across any distance.

1. In your group, write an equation that gives the number of support columns required for any given distance.

2. What other factors should be considered when determining the number of columns needed?

3. How many beams will be required for a bridge that is 1,500 feet long?

4. How many beams will be required for a bridge that is 750 feet long?

5. Think about your answer to Question 3. What recommendation would you make to the department of transportation based on that answer?

6. If the Lake Pontchartrain bridge is approximately 24 miles long, what is the minimum number of beams that would be required to span this length?

Name: ______________________________

STUDENT HANDOUT, PAGE 1

ICE WEDGING ACTIVITY

MATERIALS (PER GROUP)

- 1 balloon
- Water
- 2 half-gallon milk cartons
- Permanent marker
- Plaster of paris

INSTRUCTIONS

1. Mark each milk carton with your group name and label them 1 and 2.
2. Fill the balloon with water. Be sure that the balloon will fit into the milk carton.
3. Measure the circumference of your balloon and record it here: _______ cm
4. Measure the mass of your balloon and record it here: _______ gm
5. Following your teacher's instructions for how much water to use, mix up enough plaster of paris so that you will be able to fill up each milk carton approximately halfway. The plaster of paris will be your "rock."
6. In milk carton 1, fill ¹/₃ of the carton with plaster of paris, then place the water-filled balloon inside. Cover the balloon with plaster of paris.
7. In milk carton 2, fill the carton halfway with plaster of paris.
8. Allow both cartons to dry, and then place them in the freezer overnight.
9. Predict what you think will happen to the plaster of paris in carton 1:

10. Predict what you think will happen to the plaster of paris in carton 2:

Name: ______________________________

STUDENT HANDOUT, PAGE 2

ICE WEDGING ACTIVITY

11. After the cartons have frozen, peel away the milk carton from the frozen plaster of paris.
12. Observe the plaster of paris blocks and record observations for both cartons:
    Carton 1:

    Carton 2:

13. Next, remove the water-filled balloon from carton 1. If you are unable to remove it with your hands, ask your teacher to use a chisel and a hammer to gently remove the plaster from the balloon.
14. Measure the circumference of the balloon: _______ cm
15. Measure the mass of the balloon: _______ gm
16. Compare the circumference and mass of your balloon after it was frozen to before it was frozen. How much did the circumference change? _______ cm
    How much did the mass change? _______ gm
17. How do you think these changes in the balloon affected your plaster of paris "rock"?

IMAGES OF ROAD WEATHERING

Pothole

Cracked asphalt

Road erosion

Tree root growing under road

Rock Observation Rubric

Name: ______________________________

Criteria	Below Standard (1 point)	Approaching Standard (2 points)	Meets or Exceeds Standard (3 points)	*Self or Peer Assessment*	*Teacher Assessment*
WRITTEN ROCK OBSERVATIONS *Include for every rock sample*	Many rock samples are missing at least three characteristics.	Some mystery rocks are missing three written characteristics or student didn't make observations for all rocks.	At least three observations are documented for each rock sample.		
SKETCH OF ROCK SAMPLE *Include for each rock and include labels to explain the rocks' characteristics*	Sketches or labels are missing for three or more rocks.	Sketches or labels are missing for one or two rocks.	Labeled sketches are provided for all rock samples.		
PROFESSIONALISM	There are many spelling, punctuation, and grammar errors and many incomplete sentences.	There are some spelling, punctuation, and grammar errors and some incomplete sentences.	Proper spelling, punctuation, and grammar are used throughout observations. (This includes using complete sentences.)		

TOTAL SCORE (SELF OR PEER): ________

TOTAL SCORE (TEACHER): ________

COMMENTS:

Social Studies Bridge Multimedia Presentation Rubric Name: ______________				
Criteria	Below Standard (1 point)	Approaching Standard (2 points)	Meets or Exceeds Standard (3 points)	*Score*
HISTORICAL SIGNIFICANCE	Students do not describe the historical significance of their bridge.	Students provide some accurate historical information but do not describe the bridge's historical significance.	Students provide accurate historical information and describe the bridge's historical significance.	
BASIC FACTS	Students provide little information about their bridge.	Students provide some basic facts but do not provide enough detail on their bridge.	Students give age, materials used, and type and describe the overall structure of their bridge.	
FUNCTION	Students fail to describe the usage and special features of their bridge.	Students provide some description about the function of their bridge, but fail to describe how their bridge affects the community in which it is found.	Students adequately describe the function and purpose of their bridge. This may include usage, impact on community, and special features.	
OVERALL PRESENTATION	Not all team members participated.	All team members participate in the presentation; however, presentation is disorganized and difficult to follow.	Presentation is clear, organized, and engaging. It is clear that all students participated in the development and presentation of the project.	
TOTAL SCORE: ________ COMMENTS:				

Lesson Plan 3: Arch Bridges

In this lesson, students begin to explore arch bridges. Through experimentation, students investigate strength, span length, and overall structure of arch bridges. Using this information, students investigate possible construction sites and determine whether an arch bridge would be an appropriate bridge design for that location. Finally, students explore what role arches have played in the development of infrastructure throughout history.

ESSENTIAL QUESTIONS

- What are the strengths and limitations of an arch bridge?
- How do beam and arch bridges compare?
- What role have arches played in the development of infrastructure across time and culture?
- How are different types of rocks formed?

ESTABLISHED GOALS AND OBJECTIVES

At the conclusion of this lesson, students will be able to do the following:

- Collect and organize data through experimentation.
- Interpret data and write a linear equation that best fits the data.
- Use a linear model to solve problems in a real-world context.
- Develop a game that accurately teaches the rock cycle to elementary-age children.
- Understand the role that arches have played in the development of infrastructure across time and culture.

TIME REQUIRED

- 3 days (approximately 45 minutes each day; see Tables 3.7–3.8, pp. 39–40)

MATERIALS

- STEM Research Notebooks
- Handouts (attached at the end of this lesson)
- Cardstock (5 pieces per group)
- 200 pennies (per group)
- Wooden blocks (2 per group)

- Paper cup (1 per group)
- Graph paper or access to graphing software
- Safety glasses or safety goggles

SAFETY NOTES

1. All laboratory occupants must wear safety glasses or goggles during all phases of this inquiry activity.
2. Make sure all materials are put away after completing the activity.
3. Wash hands with soap and water after completing this activity.

CONTENT STANDARDS AND KEY VOCABULARY

Table 4.7 lists the content standards from the *NGSS, CCSS,* and the Framework for 21st Century Learning that this lesson addresses, and Table 4.8 (p. 113) presents the key vocabulary. Vocabulary terms are provided for both teacher and student use. Teachers may choose to introduce some or all of the terms to students.

Table 4.7. Content Standards Addressed in STEM Road Map Module Lesson 3

NEXT GENERATION SCIENCE STANDARDS

PERFORMANCE EXPECTATIONS

- MS-ESS2-1. Develop a model to describe the cycling of Earth's materials and the flow of energy that drives this process.
- MS-ESS3-1. Construct a scientific explanation based on evidence for how the uneven distributions of Earth's mineral, energy, and groundwater resources are the result of past and current geoscience processes.
- MS-ETS1-4. Develop a model to generate data for iterative testing and modification of a proposed object, tool, or process such that an optimal design can be achieved.

SCIENCE AND ENGINEERING PRACTICES

Developing and Using Models

Modeling in 6–8 builds on K–5 experiences and progresses to developing, using, and revising models to describe, test, and predict more abstract phenomena and design systems.

- Develop and use a model to describe phenomena.
- Develop a model to generate data to test ideas about designed systems, including those representing inputs and outputs.

Continued

Table 4.7. *(continued)*

Analyzing and Interpreting Data

Analyzing data in 6–8 builds on K–5 experiences and progresses to extending quantitative analysis to investigations, distinguishing between correlation and causation, and basic statistical techniques of data and error analysis.

- Analyze and interpret data to provide evidence for phenomena.

Constructing Explanations and Designing Solutions

Constructing explanations and designing solutions in 6–8 builds on K–5 experiences and progresses to include constructing explanations and designing solutions supported by multiple sources of evidence consistent with scientific ideas, principles, and theories.

- Construct a scientific explanation based on valid and reliable evidence obtained from sources (including the students' own experiments) and the assumption that theories and laws that describe the natural world operate today as they did in the past and will continue to do so in the future.

Scientific Knowledge Is Open to Revision in Light of New Evidence (Nature of Science practice)

- Science findings are frequently revised and/or reinterpreted based on new evidence.

DISCIPLINARY CORE IDEAS

ESS2.A: Earth's Materials and Systems

- All Earth processes are the result of energy flowing and matter cycling within and among the planet's systems. This energy is derived from the sun and Earth's hot interior. The energy that flows and matter that cycles produce chemical and physical changes in Earth's materials and living organisms.

ETS1.B: Developing Possible Solutions

- A solution needs to be tested, and then modified on the basis of the test results, in order to improve it.
- Models of all kinds are important for testing solutions.

ETS1.C: Optimizing the Design Solution

- The iterative process of testing the most promising solutions and modifying what is proposed on the basis of the test results leads to greater refinement and ultimately to an optimal solution.

CROSSCUTTING CONCEPTS

Stability and Change

- Explanations of stability and change in natural or designed systems can be constructed by examining the changes over time and processes at different scales, including the atomic scale.

Continued

Table 4.7. *(continued)*

Cause and Effect

- Cause and effect relationships may be used to predict phenomena in natural or designed systems.

Influence of Science, Engineering, and Technology on Society and the Natural World

- All human activity draws on natural resources and has both short- and long-term consequences, positive as well as negative, for the health of people and the natural environment.

COMMON CORE STATE STANDARDS FOR MATHEMATICS

MATHEMATICAL PRACTICES

- MP1. Make sense of problems and persevere in solving them.
- MP2. Reason abstractly and quantitatively.
- MP3. Construct viable arguments and critique the reasoning of others.
- MP4. Model with mathematics.
- MP5. Use appropriate tools strategically.
- MP6. Attend to precision.

MATHEMATICAL CONTENT

- 8.EE.B.5. Graph proportional relationships, interpreting the unit rate as the slope of the graph. Compare two different proportional relationships represented in different ways. For example, compare a distance-time graph to a distance-time equation to determine which of two moving objects has greater speed.
- 8.EE.C.7.B. Solve linear equations with rational number coefficients, including equations whose solutions require expanding expressions using the distributive property and collecting like terms.
- 8.F.B.5. Describe qualitatively the functional relationship between two quantities by analyzing a graph (e.g., where the function is increasing or decreasing, linear or nonlinear). Sketch a graph that exhibits the qualitative features of a function that has been described verbally.

COMMON CORE STATE STANDARDS FOR ENGLISH LANGUAGE ARTS

(Note: These standards arch over the entire bridge biography project and include interactions within the writing workshop.)

READING STANDARD

- RI.8.7. Evaluate the advantages and disadvantages of using different mediums (e.g., print or digital text, video, multimedia) to present a particular topic or idea.

Continued

Table 4.7. (*continued*)

WRITING STANDARDS

- W.8.1.E. Provide a concluding statement or section that follows from and supports the argument presented.
- W.8.2. Write informative/explanatory texts to examine a topic and convey ideas, concepts, and information through the selection, organization, and analysis of relevant content.
- W.8.2.A. Introduce a topic clearly, previewing what is to follow; organize ideas, concepts, and information into broader categories; include formatting (e.g., headings), graphics (e.g., charts, tables), and multimedia when useful to aiding comprehension.
- W.8.2.B. Develop the topic with relevant, well-chosen facts, definitions, concrete details, quotations, or other information and examples.
- W.8.2.C. Use appropriate and varied transitions to create cohesion and clarify the relationships among ideas and concepts.
- W.8.2.D. Use precise language and domain-specific vocabulary to inform about or explain the topic.
- W.8.2.E. Establish and maintain a formal style.
- W.8.2.F. Provide a concluding statement or section that follows from and supports the information or explanation presented.
- W.8.3. Write narratives to develop real or imagined experiences or events using effective technique, relevant descriptive details, and well-structured event sequences.
- W.8.3.A. Engage and orient the reader by establishing a context and point of view and introducing a narrator and/or characters; organize an event sequence that unfolds naturally and logically.
- W.8.3.D. Use precise words and phrases, relevant descriptive details, and sensory language to capture the action and convey experiences and events.
- W.8.6. Use technology, including the internet, to produce and publish writing and present the relationships between information and ideas efficiently as well as to interact and collaborate with others.
- W.8.8. Gather relevant information from multiple print and digital sources, using search terms effectively; assess the credibility and accuracy of each source; and quote or paraphrase the data and conclusions of others while avoiding plagiarism and following a standard format for citation.

SPEAKING AND LISTENING STANDARDS

- SL.8.1. Engage effectively in a range of collaborative discussions (one-on-one, in groups, and teacher-led) with diverse partners on grade 8 topics, texts, and issues, building on others' ideas and expressing their own clearly.
- SL.8.1.A. Come to discussions prepared, having read or researched material under study; explicitly draw on that preparation by referring to evidence on the topic, text, or issue to probe and reflect on ideas under discussion.

Continued

Table 4.7. (*continued*)

SPEAKING AND LISTENING STANDARDS (*continued*)

- SL.8.1.B. Follow rules for collegial discussions and decision making, track progress toward specific goals and deadlines, and define individual roles as needed.
- SL.8.1.C. Pose questions that connect the ideas of several speakers and respond to others' questions and comments with relevant evidence, observations, and ideas.
- SL.8.1.D. Acknowledge new information expressed by others, and, when warranted, qualify or justify their own views in light of the evidence presented.
- SL.8.3. Delineate a speaker's argument and specific claims, evaluating the soundness of the reasoning and relevance and sufficiency of the evidence and identifying when irrelevant evidence is introduced.
- SL.8.4. Present claims and findings, emphasizing salient points in a focused, coherent manner with relevant evidence, sound valid reasoning, and well-chosen details; use appropriate eye contact, adequate volume, and clear pronunciation.
- SL.8.6. Adapt speech to a variety of contexts and tasks, demonstrating command of formal English when indicated or appropriate.

LANGUAGE STANDARDS

- L.8.5. Demonstrate understanding of figurative language, word relationships, and nuances in word meanings.

FRAMEWORK FOR 21ST CENTURY LEARNING

- Interdisciplinary Themes: Civic Literacy
- Learning and Innovation Skills: Creativity and Innovation; Critical Thinking and Problem Solving; Communication and Collaboration
- Information, Media, and Technology Skills: Information Literacy; Media Literacy; ICT Literacy
- Life and Career Skills: Flexibility and Adaptability; Initiative and Self-Direction; Social and Cross-Cultural Skills; Productivity and Accountability; Leadership and Responsibility

Table 4.8. Key Vocabulary for Lesson 3

Key Vocabulary	Definition
abutments	"the outermost end supports on a bridge, which carry the load from the deck" (PBS 2001)
arch	a curved symmetrical shape
arch bridge	"a curved structure that converts the downward force of its own weight, and of any weight pressing down on top of it, into an outward force along its sides and base" (PBS 2001)

TEACHER BACKGROUND INFORMATION

Mathematics

The arch bridge is one of the earliest bridge designs. Because of its beauty and strength, it is a design that has literally stood the test of time. Many early examples of arch bridges are still standing today. Most early arch bridges were known as filled barrel arch bridges. These designs consisted of dirt and stones that filled the gap between the arch and the deck. This form of construction added strength to the overall structure. However, because of the weight of the material used, these bridges were limited in size. To span long distances, arch bridges had to be constructed using a series of arches. As building materials have improved, engineers have been able to use single arches to span much greater lengths. The increased size of these bridges has led engineers on a quest to determine what shape of arch is the strongest. Many are still working to answer this question. One theory states that fixing the ends of a chain or string and allowing it to hang can model the "ideal" arch. However, determining the equation that models such a shape is well beyond the scope of this module.

As with a beam bridge, the forces acting on an arch bridge are compression and tension. In an arch bridge, tension is minimal. Compression is the main force that acts on an arch bridge. Because of the design, the weight of the deck and fill material compresses the arch, which distributes the weight outward and into the abutments at either end of the arch. For further information about arch bridges, visit the following websites:

- *www.historyofbridges.com/facts-about-bridges/arch-bridges*
- *http://science.howstuffworks.com/engineering/civil/bridge5.htm*

Science

Rocks recycle into other types of rocks via the rock cycle. For example, igneous rocks turn into sediment via weathering and erosion. This sediment becomes a sedimentary rock through compacting and cementation of the particles. If buried and subjected to intense heat and pressure, the sedimentary rock will become metamorphic rock, and so on. Learn more about the rock cycle at the following resources:

- *www.bbc.co.uk/bitesize/ks3/science/environment_earth_universe/rock_cycle/activity*
- *www.learner.org/interactives/rockcycle/diagram.html*

ELA

For help on conducting a writing workshop in the middle school classroom and content area writing, see the following resources:

- Write in the Middle: A Workshop for Middle School Teachers: *www.learner.org/resources/series192.html*
- *Content-Area Writing: Every Teacher's Guide* by Harvey Daniels, Steven Zemelman, and Nancy Steineke (Heinemann, 2007)
- *Using the Workshop Approach in the High School English Classroom: Modeling Effective Writing, Reading, and Thinking Strategies for Student Success* by Cynthia D. Urbanski (Corwin Press, 2015)

Social Studies

Ancient civilizations such as the Egyptians, Greeks, and Babylonians all used arches in their construction. However, it was the ancient Romans who perfected the use of the arch and were able to use it to construct giant structures (Alchin 2015). Romans used arches to build and develop their infrastructure (e.g., the Coliseum, aqueducts, sewers, palaces, and temples). Through the use of concrete, Romans were able to develop arches that could hold large amounts of weight (Eduplace 2015). Other cultures adopted the use of the arch (e.g., Byzantine, Romanesque) and Muslim architects even improved on the Romans' design by developing pointed, scalloped, and horseshoe arches (Eduplace 2015). The Romans also improved on their arch design to form domes. The Pantheon is an example of a dome structure in ancient Rome. Arch structures are used throughout our world today, and we owe it all to the ancient civilizations who recognized their strength.

In social studies, students are challenged to create a game for elementary students that focuses on the rock cycle. The target grade range for the game should be grades 3–5. Kids Geology provides an example of a rock game at *www.kidsgeo.com/geology-games/rocks-game.php*. You may wish to review this, and similar games, as preparation for this activity.

COMMON MISCONCEPTIONS

Students will have various types of prior knowledge about the concepts introduced in this lesson. Table 4.9 (p. 116) outlines some common misconceptions students may have concerning these concepts. Because of the breadth of students' experiences, it is not possible to anticipate every misconception that students may bring as they approach this lesson. Incorrect or inaccurate prior understanding of concepts can influence student learning in the future, however, so it is important to be alert to misconceptions such as those presented in the table.

Table 4.9. Common Misconceptions About the Concepts in Lesson 3

Topic	Student Misconception	Explanation
Variables	There is confusion about which are the independent and dependent variables.	An independent variable is a variable that you can control (span length of a bridge). A dependent variable is a variable that you observe and measure (weight the span will hold).
Models	Models are only in 3-D.	Models in fact can be 2-D representations such as pictures, graphs, written descriptions, blueprints, equations, and other representations, as well as a 3-D model.
	A model's usefulness is based solely on the model's physical resemblance to the object being modeled.	A physical model is a smaller or larger physical copy of an object such as a bridge. The model represents a similar object in the sense that scale is an important characteristic of the model.
Force (a push or pull between objects)	Nonmoving objects do not exert a force (e.g., a tabletop or roadway).	Stationary objects can exert forces on other objects. For example, when you roll a toy car over a tabletop, there is a frictional force between the table and the car wheels.
	If a moving object is slowing, the force that was propelling it forward is decreasing.	When a force acts on a moving object in a direction opposite the direction of motion, the moving object will slow, even if the force that was propelling the object forward continues.
Weathering, erosion, and deposition (the breaking down of a material, the movement of the material to another area)	Water and wind cannot wear away rock and other building materials.	Water and wind can wear away rock and other building materials.
	Water freezing in cracks cannot break apart rock.	Water freezing in cracks can break apart rock.
Earth materials (types and uses of rocks)	Rocks have little practical use.	Rocks are used for a wide variety of consumer and industrial purposes. For example, rocks are used in paper production, cement, household cleaners, jewelry, pencils, chalk, glass, and building materials such as bricks and kitchen countertops.
	All rocks are the same.	There are three major types of rocks (sedimentary, igneous, and metamorphic), which have a variety of mineral compositions.

PREPARATION FOR LESSON 3

Gather materials for the mathematics investigation and make copies of the student handouts attached at the end of this lesson. Reserve computers or web-enabled devices for use in science, ELA, and social studies.

Contact an elementary school to arrange for grade 3–5 students to play and provide feedback on the rock games teams will create in social studies.

LEARNING COMPONENTS

Introductory Activity/Engagement

Connection to the Challenge: Begin each day of this lesson by directing students' attention to the driving question for the module and challenge: How can we develop a decision model to help us make a recommendation to the local department of transportation on the type of bridge to build for a given location? Ask students why bridges are so important and what impact a bridge might have on a community. Hold a brief student discussion of how their learning in the previous days' lesson(s) contributed to their ability to create their plan and build their prototype. You may wish to hold a class discussion, creating a class list of key ideas on chart paper, or you may wish to have students create a notebook entry with this information.

Driving Questions for Lesson 3: How are arch bridges different from beam bridges? Do they support differently? Are they found in different environments?

Mathematics Class: Show students a picture of a deep canyon (p. 144) that is more than 250 feet across and discuss the following questions:

- Would a beam bridge be appropriate for this site?
- What are some limitations of a beam bridge?
- Could we build a column or columns in the middle to support the weight of the bridge?
- Would it be cost effective to build these columns?
- What characteristics might the bridge across this canyon need?

Science Connection: Pose the following question to students: Where do different types of rocks come from? Have students discuss their answers to the question in small groups, then have them share their answers with the whole class. Record students' ideas. If students are having trouble recalling from previous grades how different rocks form, you could pose questions such as the following:

- What are sedimentary rocks composed of? (*tiny pieces or fragments of rocks that have been cemented together*)
- How did the rock pieces get so small? (*weathering or erosion*)
- What is the source of igneous rocks? (*cooled magma or lava*)
- How do metamorphic rocks come to be? (*placing existing rocks under extreme heat and pressure*)

Inform the class that a local elementary school (or an afterschool program for elementary students) is in need of ways to help teach the elementary children about minerals, rocks, and the rock cycle in fun and engaging ways. This class is going to work in teams to design a game appropriate for elementary students in grades 3–5. The game can be a computer game, a board game, a role-playing game, or another creative (but teacher-approved) option.

ELA Connection: Launch a writing workshop for the bridge biographies. Introduce procedures for collaborating, creating, accessing information and resources, and engaging in peer and teacher conferences. (See the Bridge Biography Composition and Production Proposal handout on p. 140.)

Social Studies Connection: Ask students: What is an arch? What are some different contexts in which you have seen arches? How are they used?

Activity/Exploration

Mathematics Class: Put students with their project groups and provide them with the following problem, which also can be found on the Arch Bridge Weight Test handout (p. 127): *The local department of transportation is once again enlisting our help. For this investigation, we will be exploring the strength of arch bridges. As before, the department wants us to determine how much building material to use. However, this time, due to the length of the bridge and the amount of traffic it must support, your bridge will need to hold 200 pennies!*

Review the procedures listed on the handout and ensure each group has the proper setup. Then, have students begin to collect data. Possible facilitating questions to ask while they are working include the following:

- How is this structure different from the beam bridge?
- How is the structure of the arch bridge affected as you add weight?
- Can you describe how the arch bridge fails? How is this different from the beam bridge?

- What are the variables involved in this problem? Why are we not changing the thickness of the deck? If we were to change the deck thickness also, what would that do to our ability to describe what is happening with the arch?
- How can we organize our data so we can *see* what is happening as we increase the thickness of our arch?

STEM Research Notebook Prompt

Have students reflect in their STEM Research Notebooks on the following:

- *Explain why a beam bridge might not be a good choice across a wide, deep canyon.*
- *Explain how the structure of an arch bridge differs from a beam bridge.*

Science Connection: Give students the Rock Cycle Game Planning Guide handout (p. 124). Students will use what they learned about rocks and minerals from Lessons 1 and 2 to help create a rock cycle diagram and develop the game.

Before students create their diagrams, have them explore the animations on the following websites about the rock cycle:

- *www.learner.org/interactives/rockcycle/change.html*
- *www.windows2universe.org/earth/geology/rocks_intro.html*

ELA Connection: Introduce the bridge biography writing workshop. Discuss with students how the process of a writing workshop is similar to the teamwork of engineers. The idea here is for students to realize that teamwork, collaboration, and creation are similar to what happens during a writing workshop.

Social Studies Connection: Have students research and develop a poster or another type of presentation that explains the role arches have played in the development of infrastructure throughout history.

Explanation

Mathematics Class: If students have not finished their investigation, give students the opportunity to work with their groups to finish the investigation questions on the Arch Bridge Weight Test handout. Once students have completed the investigation, bring them together for a group discussion and pose the following questions:

- What variables did your group consider when trying to solve the problem?
- How did your group decide to display your data? Why?

- Can you describe the relationships you saw when you displayed your data?
- What equation did you write to help you model this situation? How did you find it?
- Did anyone else come up with a different equation?
- Why are they not the same? What does this mean about models? What should we keep in mind when we extrapolate as we try to determine how strong something will be based on a model?

Give each student an Arch Bridge Basics handout (p. 132). Discuss as a class each term and have students label the diagram provided on the handout. As you discuss the vocabulary, spend a few minutes talking about the abutments. Because of the structure of an arch bridge, nearly all of the weight is distributed around the arch and to the abutments. Allow the students to return to their groups to discuss the following questions:

- What happened as the arch got stronger and we added more weight to the bridge?
- Did anyone's wooden blocks get pushed outward during their experiment?
- How would engineers use what they know about the way weight is distributed in arch bridges as they survey sites and design arch bridges?

After a few minutes, gather the class together to discuss the students' ideas. During this discussion, make sure to discuss the forces involved with arch bridges. It is important to help students recognize that compression is the main force acting on an arch bridge. While tension forces are present, they are in most cases negligible. This is what allowed arch bridges to be constructed from materials with little tensile strength such as cast iron and other early building materials.

STEM Research Notebook Prompt

Have students reflect on the following prompt in their STEM Research Notebooks: *What would an engineer need to consider if community members suggested they wanted an arch bridge built at a particular location in their town?*

Science Connection: Groups compare their rock cycle diagrams with a neighboring group and check for similarities and differences among their diagrams. Student groups will reconcile their differences and modify their rock cycle diagrams. After groups have had a chance to modify their diagrams, they will share with the whole class. The first group will present their diagram to the class. (This can be done with a document viewer, projected on an interactive white board, or drawn on a standard white board.) Subsequent groups will modify the rock cycle diagram from the previous group. Facilitate the groups' discussions of their rock cycle diagrams and make sure that all parts of the rock cycle are presented.

STEM Research Notebook Prompt

Ask students to discuss in a STEM Research Notebook entry why it is important for an engineer to understand the rock cycle and how this might connect to designing a bridge. You could also have students reflect on the statement "Engineers learn about the natural world to design the human-built environment." Have students record their ideas in the STEM Research Notebook.

ELA Connection: Conduct the bridge biography writing workshop.

Social Studies Connection: Students share their posters or presentations with their peers. Bring students back together after each group has had an opportunity to share its findings. Have a class discussion about the posters and presentations, and have students do a gallery walk to see all of the posters and presentations. Have student reflect on when and how arches were used throughout history.

Elaboration/Application of Knowledge

Mathematics Class: Place students in their project groups and have them answer the following questions to reactivate what has been discussed about arch bridges:

- What are the key features of an arch bridge and what function does it serve?
- What sites do you believe are appropriate for an arch bridge?
- Describe the forces that are present in an arch bridge. Draw a picture that shows these forces.

Give the students the Arch Bridge—Span Length handout (p. 133). Ask students to make observations about the similarities of the bridges in the pictures shown and ask them to make a prediction about which bridge they believe is the strongest. During this discussion, help students think about that even though the first two bridges are made of rock and stone, the "fill" makes them very strong. Encourage students to deepen their understanding by discussing the following questions in their groups:

- What would happen with the first two bridges as we attempt to increase the size?
- Why do you think steel allows us to build larger bridges?
- What characteristic of the third bridge helps evenly distribute the weight of the deck across the entire arch?

Ask the students to work through the span length investigation. The goal of this investigation is for students to develop models that give dimensions of an arch bridge as the span length increases. Possible models include span length vs. height of the arch and

span length vs. length of the arch. As the students work in their groups, circulate around the room and ask the following questions:

- What are some different measurements we could take from the diagram?
- How are these measurements related to the span length?
- How did your group determine the length of the arch?
- Can you write a model that relates these measurements?
- How can we use those models to describe an arch bridge as the span length increases?

Once groups have finished their investigation, bring the class together to share their findings.

STEM Research Notebook Prompt

Have students reflect on the pros and cons of an arch bridge. Then have students complete the following statement in their STEM Research Notebooks: *An arch bridge is the best design choice when …*

Science Connection: In small groups, students design their rock cycle game to teach the rock cycle to elementary-age students. (It would be ideal if the students could take their completed games to an elementary classroom to share with real students.) Students will use the rock cycle diagram that they created as a class to develop the rules of the game. The game should be fun, but it should also be scientifically accurate.

ELA Connection: Conduct the bridge biography writing workshop.

Social Studies Connection: Not applicable.

Evaluation/Assessment

Students may be assessed on the following performance tasks and other measures listed.

Performance Tasks

- End of Lesson Assessment: Have students use a Venn diagram to compare and contrast beam and arch bridges.
- Share game with elementary schools (if possible)
- Arches in History Poster Checklist
- Rock Cycle Game Rubric and Peer Assessment Rubric
- Bridge Biography Rubric

Other Measures

- STEM Research Notebook Entries. You should regularly read and respond to students in their STEM Research Notebooks. Your response should not indicate whether students' entries are right or wrong. Instead, include comments or questions that will push and stretch students' thinking and can aid students in moving toward development of their decision models.
- Learning activity responses: Arch Bridge Weight Test, Arch Bridge Basics, and Arch Bridge—Span Length handouts.

INTERNET RESOURCES

Information about arch bridges

- *www.historyofbridges.com/facts-about-bridges/arch-bridges*
- *http://science.howstuffworks.com/engineering/civil/bridge5.htm*

Rock cycle resources

- *www.bbc.co.uk/bitesize/ks3/science/environment_earth_universe/rock_cycle/activity*
- *www.learner.org/interactives/rockcycle/diagram.html*
- *www.learner.org/interactives/rockcycle/change.html*
- *www.windows2universe.org/earth/geology/rocks_intro.html*

Information on conducting a middle school writing workshop

- *www.learner.org/resources/series192.html*

Example of a rock game

- *www.kidsgeo.com/geology-games/rocks-game.php*

Name: ______________________________

STUDENT HANDOUT

ROCK CYCLE GAME PLANNING GUIDE

OBJECTIVE

Your team will develop a game that accurately depicts and teaches the rock cycle to elementary students.

RULES

1. Your team can decide the format of your game. For example, it can be a traditional board game, a role player game, or an online game. Your only limitation is your team's creativity. Also, remember that your team has a deadline. This assignment is due on ____________________. It is a good idea to share your idea with your teacher before you begin.
2. Your game must accurately teach the rock cycle.
3. Your game must be simple enough that elementary students can easily understand and follow the rules.
4. On the due date above, you will present your game to the class. A group of your peers will play your game and assess it using the Peer Assessment Rubric (see separate handout). You will do the same to a game that your peers create.
5. Have fun, be creative, and make an awesome game that teaches the rock cycle!

ASSESSMENT

See the attached assessment rubrics.

Name: ____

Rock Cycle Game Rubric

The following scale will be used to rate the criteria for the game:

1 = Not at all 2 = Somewhat 3 = For the most part 4 = YES!

Category	Criteria	*Score*
EFFORT	There is evidence that excellent effort was put into creating the game. Evidence includes a well-organized presentation and a complete, functioning game.	
VOCABULARY	The game accurately teaches terminology that elementary students will need to understand and describe the rock cycle.	
ACCURACY	The game accurately teaches the rock cycle.	
PLAYABILITY	Elementary students will be able to play the game. The rules of the game are clearly outlined and easy to follow.	
PEER ASSESSMENT	This is an average score from your peers (using the average of the scores from the Peer Assessment Rubrics).	
TEAM CONTRIBUTION	All members contributed to the development and creation of the game.	
TOTAL SCORE: _______ COMMENTS:		

Name: ____

STUDENT HANDOUT

ROCK CYCLE GAME PEER ASSESSMENT RUBRIC

Name of game: ____

Who created the game? ________________________________

Who played the game? _________________________________

On a scale of 1–4, rate the following four categories for the game you played.

1 = Not at all **2 = Somewhat** **3 = For the most part** **4 = YES!**

Category	Criteria	Score
Rules	Were the rules clearly outlined and easy to understand?	
Accuracy	Was the rock cycle information accurate and complete?	
Effort	Was there effort put into creating this game? Were all the game pieces included (if applicable)?	
Fun!	Did you have fun playing this game?	
	Total Score	

Name: ______________________________

STUDENT HANDOUT, PAGE 1

ARCH BRIDGE WEIGHT TEST

PROBLEM

The local department of transportation is once again enlisting our help. For this investigation, we will be exploring the strength of arch bridges. As before, the department wants us to determine how much building material to use. However, this time, due to the length of the bridge and the amount of traffic it must support, your bridge will need to hold 200 pennies!

MATERIALS

- Cardstock: As before, you are being given cardstock that has been engineered to match the building materials that will be used in the construction of the bridge. However, because of the high cost of the cardstock building material, you will only be given five sheets!
- 200 pennies: The department has determined that this is proportional to the amount of weight that the real bridge will need to support.
- Wooden blocks
- Paper cup
- Graph paper or access to graphing software
- Safety glasses or safety goggles

SAFETY NOTES

1. All laboratory occupants must wear safety glasses or goggles during all phases of this inquiry activity.
2. Make sure all materials are put away after completing the activity.
3. Wash hands with soap and water after completing this activity.

SETUP

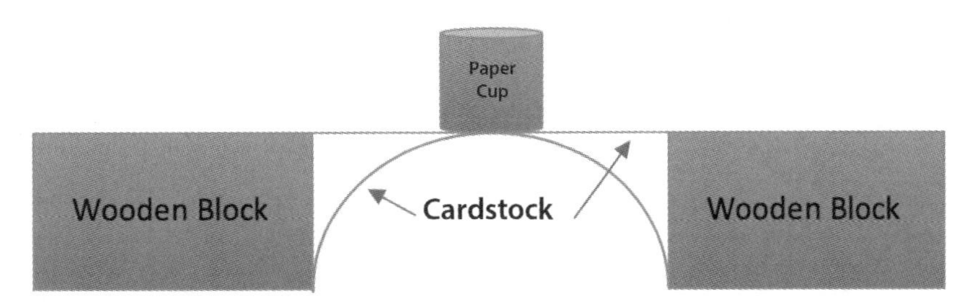

Name: ____

STUDENT HANDOUT, PAGE 2

ARCH BRIDGE WEIGHT TEST

PROCEDURE

1. Use the diagram above to set up the test site. Be sure to place the objects in the labeled positions.
2. Place a paper cup in the middle of your bridge. This will allow you focus where you want to place your weight without worrying about the pennies falling off.
3. Test your bridge as the arch increases in thickness. To make the bridge stronger, we will add a layer of cardstock to the arch. The deck will remain one sheet thick through the entire investigation. Make sure to add one penny at a time. As you are adding weight, be sure to observe how the bridge behaves. How does it fail? Are there weak points? Is it a gradual fail?
4. Record and interpret data.

Layers of Cardstock	**Number of Pennies *Before* Collapse**	**Observations**

INVESTIGATION QUESTIONS

1. What are the variables involved in this problem?

Name: ____

STUDENT HANDOUT, PAGE 3

ARCH BRIDGE WEIGHT TEST

2. How can you organize the data to give a clearer picture of what is happening?

3. Describe the relationship between your variables.

4. How could you construct a model (equation) that relates the variables?

5. With your group, determine how many layers of concrete (cardstock) you must use to make your arch support the required amount of weight.

6. Use the equation your group found in this investigation and compare it with the model you found for the strength of a beam bridge. How are they the same? How are they different? What do these differences mean?

Name: ______________________________

ARCH BRIDGE WEIGHT TEST

7. How are these differences and similarities displayed on your representation of the data?

8. What did you notice about the wooden blocks as you added weight to the bridge? What does this tell you about how the arch distributes the weight on the bridge?

9. What does this mean for city planners as they try to determine whether an arch bridge is a suitable design for a particular site?

10. Thinking about how the bridge failed, what are some ways, besides making the arch thicker, that you could strengthen an arch bridge?

Name: ______________________________________

STUDENT HANDOUT, PAGE 5

ARCH BRIDGE WEIGHT TEST

If graphing technology (e.g., graphing calculators, online graphing calculators, GeoGebra, Desmos) is available, then answer the following questions:

11. Are the data perfectly linear?

12. Use the calculator to find a model that might be a "better" fit for the data.

Name: ______________________________

STUDENT HANDOUT

ARCH BRIDGE BASICS

Write the definition of the following terms. Be sure to label your picture.

Arch:

Abutments:

Deck:

Write the definition of the following forces. Describe how these forces act on arch bridges differently than they act on a beam bridge.

Compression:

Tension:

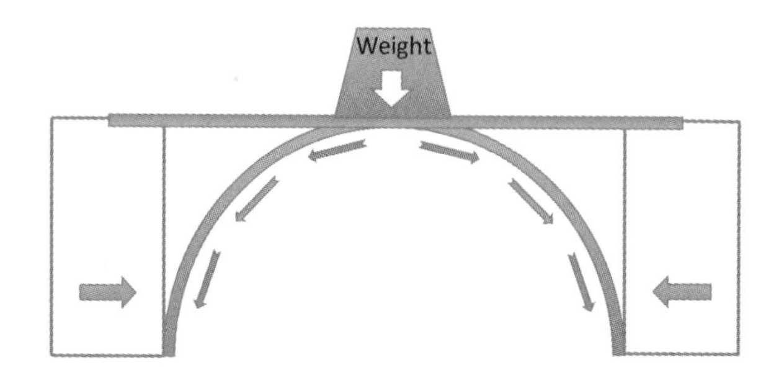

Name: ______________________________

STUDENT HANDOUT, PAGE 1

ARCH BRIDGE—SPAN LENGTH

PROBLEM

The arch bridge is one of the earliest bridge designs. Because of its beauty and strength, it is a design that has literally stood the test of time. Many early examples of arch bridges are still standing today. Most early arch bridges were known as filled barrel arch bridges. These designs consisted of dirt and stones that filled the gap between the arch and the deck (Picture 1). This form of construction added strength to the overall structure. However, because of the weight of the material used, these bridges were limited in size. To span long distances, bridges had to be constructed using a series of arches (Picture 2). As building materials have improved, engineers have been able to use single arches to span much greater lengths (Picture 3). The increased size of these bridges has led engineers on a quest to determine what shape of arch is the strongest, cheapest to build, and most appropriate for a given site.

Picture 1

Picture 2

Picture 3

Name: ____

STUDENT HANDOUT, PAGE 2

ARCH BRIDGE—SPAN LENGTH

Below is an image of the model that the city's civil engineers have used to determine what shape of arch is the strongest. They have now asked us to help create models (equations) to allow them to construct this shape of an arch at locations with varying span lengths.

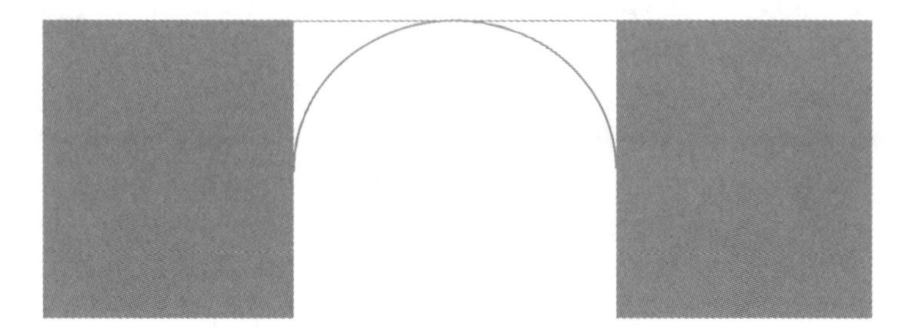

PROCEDURE

1. Using the diagram above, create models that will help the city's civil engineers determine whether this shape of arch bridge is suitable for a given site.

2. Consider the following questions:
   a. How large will the arch be as span length increases and decreases?
   b. How much material might be needed to construct the arch as span length increases?
   c. What about the site do we need to consider when attempting to build this arch?

3. Use your models to answer the investigation questions.

Name: ____

STUDENT HANDOUT, PAGE 3

ARCH BRIDGE—SPAN LENGTH

INVESTIGATION QUESTIONS

1. What is the span length shown on the diagram?

2. What is the height of the arch? Why might this measurement be important when determining if an arch bridge is an appropriate design for a particular site?

3. How long is the arch shown in the diagram? Please write a short description of how your group measured this distance. How do you know it is accurate?

4. Why is the total length of the arch important to know?

Name: ___________________________________

STUDENT HANDOUT, PAGE 4

ARCH BRIDGE—SPAN LENGTH

5. Write a model or models that help the city's civil engineers understand how the bridge features in Questions 1–4 are related.

6. Civil engineers are trying to decide whether the site in the picture below is appropriate for an arch bridge.

Name: ___________________________________

STUDENT HANDOUT, PAGE 5

ARCH BRIDGE—SPAN LENGTH

a. Use the models to help them determine how many feet of steel will be needed and how tall the arch will need to be. The distance between the tops of the two cliffs is 480 feet. The depth of the gorge is 400 feet.

b. The walls of the cliff are solid rock. Do you believe an arch bridge would be appropriate? Explain your reasoning.

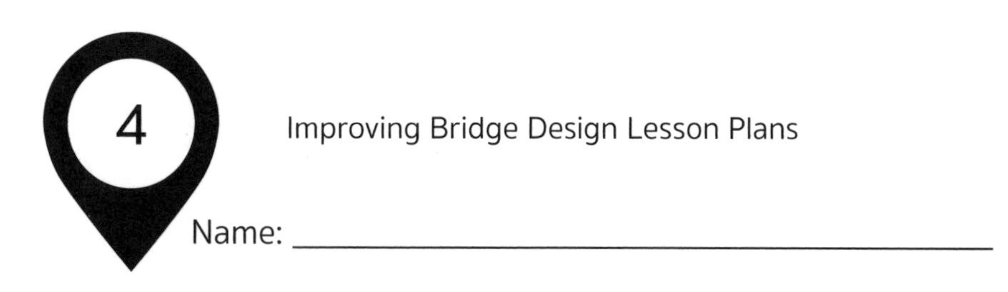

Name: ______________________________

STUDENT HANDOUT, PAGE 6

ARCH BRIDGE—SPAN LENGTH

7. Civil engineers have discovered another site. The bridge would need to span a length 400 feet over a gorge with a river that would be approximately 100 feet below the deck of the bridge. Several of the engineers are concerned that an arch bridge may not be the best option for this site. Draw a picture of the site and then use your models to help the engineers make their decision.

8. Is there any other information your group might need for a more informed answer to Questions 6 and 7?

Name: ______________________________

ARCHES IN HISTORY POSTER CHECKLIST

Student teams should research and develop a poster or a presentation that explains the role arches have played in the development of infrastructure throughout history.

Items	Yes or No	Comments
Poster/presentation is well organized.		
Poster/presentation connects arches to infrastructure.		
Poster/presentation looks at arches and infrastructure across different cultures and time periods.		
Students use graphics/ images to make poster/ presentation more engaging.		

Name: ______________________________

STUDENT HANDOUT, PAGE 1

BRIDGE BIOGRAPHY COMPOSITION AND PRODUCTION PROPOSAL

1. I am telling the life story of

2. I chose this bridge because

3. I have researched this topic in the following ways:

4. Please list, in order, important events you will highlight in the life of this bridge. Make sure you include the beginning of the bridge's "life" as well as significant changes and events over the course of time. For each event listed, note the importance of the event or change to the people in whose lives this bridge played or plays a part. *Use the back of this handout or a separate sheet of paper to create your list.*

5. Write an overall statement about the significance of this bridge in the lives of people in history.

Name: ____________________________________

STUDENT HANDOUT, PAGE 2

BRIDGE BIOGRAPHY COMPOSITION AND PRODUCTION PROPOSAL

6. Who is your proposed audience? Are you writing for children? Teens and adults? How might people use your finished product?

7. What media will you employ to tell the story of your bridge? Why did you choose these forms rather than others? How, specifically, will these media interact in the telling of your story?

8. To make my final product appropriate and highly appealing for this audience, I plan to

9. To ensure accuracy and reliability of my information, I plan to

Name: ______________________________________

STUDENT HANDOUT, PAGE 3

BRIDGE BIOGRAPHY COMPOSITION AND PRODUCTION PROPOSAL

10. I know my plans may change as I write, but right now I think I will use the following nonfiction text features and structures:

11. I believe these features and structures are particularly suited to this topic and this audience because

12. I have approximately _________ weeks to produce a final product that shows my best nonfiction writing, has appeal to my audience, and makes me proud to be an author. I understand that writing is a process and that I will learn about writing and teaching writing as I engage thoughtfully in this process. To achieve these goals, my work plan is as follows:

13. I may need help with

Name: ______________________________________

STUDENT HANDOUT, PAGE 4

BRIDGE BIOGRAPHY COMPOSITION AND PRODUCTION PROPOSAL

14. If I do need help, I plan to get it by

15. Aspects of this project that excite me are

DEEP CANYON IMAGE

Bridge Biography Rubric				
Name: ______________________				
CONTENT	Scale: No = 1; Marginally = 2; Mostly = 3; Yes = 4			
Contains accurate, clearly cited information.	No	Marginally	Mostly	Yes
The biography is comprehensive; that is, it covers the bridge's life span.	No	Marginally	Mostly	Yes
Events or people selected are relevant and significant in the lives of the bridge and community.	No	Marginally	Mostly	Yes
Clearly portrays important events and people.	No	Marginally	Mostly	Yes
Uses appropriate terminology (especially terms addressed within the module in different content areas).	No	Marginally	Mostly	Yes
WRITING PROCESS	Scale: No = 1; Marginally = 2; Mostly = 3; Yes = 4			
Student actively and productively engaged in writing workshop.	No	Marginally	Mostly	Yes
Clear evidence of effort in research and preparation.	No	Marginally	Mostly	Yes
Participated fully in presentation according to classroom parameters.	No	Marginally	Mostly	Yes
WRITING PRODUCT	Scale: No = 1; Marginally = 2; Mostly = 3; Yes = 4			
Includes an engaging introduction that establishes the context.	No	Marginally	Mostly	Yes
Clearly and logically structured, employing at least one nonfiction text structure.	No	Marginally	Mostly	Yes
Effectively employs at least three nonfiction text features to support reader comprehension.	No	Marginally	Mostly	Yes
Final product is polished and visually appealing.	No	Marginally	Mostly	Yes
TOTAL SCORE: ________				
COMMENTS:				

Lesson Plan 4: Suspension and Cable-Stayed Bridges

In this lesson, students explore the last two bridge designs, the suspension bridge and the cable-stayed bridge. While these bridges both use cables to support the road deck, they do so in different ways. Through research and experimentation, students compare and contrast the characteristics of these designs and begin to determine their ideal locations.

ESSENTIAL QUESTIONS

- What are the strengths and limitations of a suspension and cable-stayed bridge?
- What are the similarities and differences between the two types of bridges?
- How do bridges affect people's lives?
- What do road and bridge building and geology have in common?
- Why might a geologist be interested when a road construction crew uncovers the earth below?

ESTABLISHED GOALS AND OBJECTIVES

At the conclusion of this lesson, students will be able to do the following:

- Collect and organize data through experimentation.
- Interpret data and write a linear equation that best fits the data.
- Use a linear model to solve problems in a real-world context.
- Use the Pythagorean theorem to solve real-world problems.
- Explore how the building of the Waco Suspension Bridge affected people's lives on both a micro (individual/family) and macro (society) level.

TIME REQUIRED

- 3 days (approximately 45 minutes each day; see Table 3.8, p. 40)

MATERIALS

- STEM Research Notebooks
- Handouts (attached at the end of this lesson)
- Clear tape (1 roll per group)
- 1 paper clip (per group)

- Drinking straws (4 per group)
- String (approximately 2 feet per group)
- 2 oz. plastic cups (3 per group)
- Pennies (100 per group; similar weights may be used if pennies are not an option)
- Safety glasses or safety goggles

SAFETY NOTES

1. All laboratory occupants must wear safety glasses or goggles during all phases of this inquiry activity.
2. Use caution when working with sharps (e.g., scissors) to avoid cutting or puncturing skin.
3. Make sure all materials are put away after completing the activity.
4. Wash hands with soap and water after completing this activity.

CONTENT STANDARDS AND KEY VOCABULARY

Table 4.10 lists the content standards from the *NGSS, CCSS,* and the Framework for 21st Century Learning that this lesson addresses, and Table 4.11 (p. 151) presents the key vocabulary. Vocabulary terms are provided for both teacher and student use. Teachers may choose to introduce some or all of the terms to students.

Table 4.10. Content Standards Addressed in STEM Road Map Module Lesson 4

NEXT GENERATION SCIENCE STANDARDS
PERFORMANCE EXPECTATION
• MS-ETS1-4. Develop a model to generate data for iterative testing and modification of a proposed object, tool, or process such that an optimal design can be achieved.
SCIENCE AND ENGINEERING PRACTICES
Developing and Using Models
Modeling in 6–8 builds on K–5 experiences and progresses to developing, using, and revising models to describe, test, and predict more abstract phenomena and design systems.
• Develop and use a model to describe phenomena.
• Develop a model to generate data to test ideas about designed systems, including those representing inputs and outputs.

Continued

Table 4.10. (*continued*)

Analyzing and Interpreting Data

Analyzing data in 6–8 builds on K–5 experiences and progresses to extending quantitative analysis to investigations, distinguishing between correlation and causation, and basic statistical techniques of data and error analysis.

- Analyze and interpret data to provide evidence for phenomena.

Constructing Explanations and Designing Solutions

Constructing explanations and designing solutions in 6–8 builds on K–5 experiences and progresses to include constructing explanations and designing solutions supported by multiple sources of evidence consistent with scientific ideas, principles, and theories.

- Construct a scientific explanation based on valid and reliable evidence obtained from sources (including the students' own experiments) and the assumption that theories and laws that describe the natural world operate today as they did in the past and will continue to do so in the future.

Scientific Knowledge Is Open to Revision in Light of New Evidence (Nature of Science practice)

- Science findings are frequently revised and/or reinterpreted based on new evidence.

DISCIPLINARY CORE IDEA

ESS2.A: Earth's Materials and Systems

- All Earth processes are the result of energy flowing and matter cycling within and among the planet's systems. This energy is derived from the sun and Earth's hot interior. The energy that flows and matter that cycles produce chemical and physical changes in Earth's materials and living organisms.

CROSSCUTTING CONCEPTS

Stability and Change

- Explanations of stability and change in natural or designed systems can be constructed by examining the changes over time and processes at different scales, including the atomic scale.

Influence of Science, Engineering, and Technology on Society and the Natural World

- All human activity draws on natural resources and has both short- and long-term consequences, positive as well as negative, for the health of people and the natural environment.

COMMON CORE STATE STANDARDS FOR MATHEMATICS

MATHEMATICAL PRACTICES

- MP1. Make sense of problems and persevere in solving them.
- MP2. Reason abstractly and quantitatively.
- MP3. Construct viable arguments and critique the reasoning of others.

Continued

Table 4.10. *(continued)*

MATHEMATICAL PRACTICES *(continued)*

- MP4. Model with mathematics.
- MP5. Use appropriate tools strategically.
- MP6. Attend to precision.
- MP7. Look for and make use of structure.
- MP8. Look for and express regularity in repeated reasoning.

MATHEMATICAL CONTENT

- 8.EE.B.5. Graph proportional relationships, interpreting the unit rate as the slope of the graph. Compare two different proportional relationships represented in different ways. For example, compare a distance-time graph to a distance-time equation to determine which of two moving objects has greater speed.
- 8.EE.C.7.B. Solve linear equations with rational number coefficients, including equations whose solutions require expanding expressions using the distributive property and collecting like terms.
- 8.F.B.5. Describe qualitatively the functional relationship between two quantities by analyzing a graph (e.g., where the function is increasing or decreasing, linear or nonlinear). Sketch a graph that exhibits the qualitative features of a function that has been described verbally.
- 8.G.B.7. Apply the Pythagorean theorem to determine unknown side lengths in right triangles in real-world and mathematical problems in two and three dimensions.

COMMON CORE STATE STANDARDS FOR ENGLISH LANGUAGE ARTS

(Note: These standards arch over the entire bridge biography project and include interactions within the writing workshop.)

READING STANDARD

- RI.8.7. Evaluate the advantages and disadvantages of using different mediums (e.g., print or digital text, video, multimedia) to present a particular topic or idea.

WRITING STANDARDS

- W.8.1.E. Provide a concluding statement or section that follows from and supports the argument presented.
- W.8.2. Write informative/explanatory texts to examine a topic and convey ideas, concepts, and information through the selection, organization, and analysis of relevant content.
- W.8.2.A. Introduce a topic clearly, previewing what is to follow; organize ideas, concepts, and information into broader categories; include formatting (e.g., headings), graphics (e.g., charts, tables), and multimedia when useful to aiding comprehension.

Continued

Table 4.10. (*continued*)

- W.8.2.B. Develop the topic with relevant, well-chosen facts, definitions, concrete details, quotations, or other information and examples.
- W.8.2.C. Use appropriate and varied transitions to create cohesion and clarify the relationships among ideas and concepts.
- W.8.2.D. Use precise language and domain-specific vocabulary to inform about or explain the topic.
- W.8.2.E. Establish and maintain a formal style.
- W.8.2.F. Provide a concluding statement or section that follows from and supports the information or explanation presented.
- W.8.3. Write narratives to develop real or imagined experiences or events using effective technique, relevant descriptive details, and well-structured event sequences.
- W.8.3.A. Engage and orient the reader by establishing a context and point of view and introducing a narrator and/or characters; organize an event sequence that unfolds naturally and logically.
- W.8.3.D. Use precise words and phrases, relevant descriptive details, and sensory language to capture the action and convey experiences and events.
- W.8.6. Use technology, including the internet, to produce and publish writing and present the relationships between information and ideas efficiently as well as to interact and collaborate with others.
- W.8.8. Gather relevant information from multiple print and digital sources, using search terms effectively; assess the credibility and accuracy of each source; and quote or paraphrase the data and conclusions of others while avoiding plagiarism and following a standard format for citation.

SPEAKING AND LISTENING STANDARDS

- SL.8.1. Engage effectively in a range of collaborative discussions (one-on-one, in groups, and teacher-led) with diverse partners on grade 8 topics, texts, and issues, building on others' ideas and expressing their own clearly.
- SL.8.1.A. Come to discussions prepared, having read or researched material under study; explicitly draw on that preparation by referring to evidence on the topic, text, or issue to probe and reflect on ideas under discussion.
- SL.8.1.B. Follow rules for collegial discussions and decision making, track progress toward specific goals and deadlines, and define individual roles as needed.
- SL.8.1.C. Pose questions that connect the ideas of several speakers and respond to others' questions and comments with relevant evidence, observations, and ideas.
- SL.8.1.D. Acknowledge new information expressed by others, and, when warranted, qualify or justify their own views in light of the evidence presented.

Continued

Table 4.10. *(continued)*

- SL.8.4. Present claims and findings, emphasizing salient points in a focused, coherent manner with relevant evidence, sound valid reasoning, and well-chosen details; use appropriate eye contact, adequate volume, and clear pronunciation.
- SL.8.5. Integrate multimedia and visual displays into presentations to clarify information, strengthen claims and evidence, and add interest.
- SL.8.6. Adapt speech to a variety of contexts and tasks, demonstrating command of formal English when indicated or appropriate.

FRAMEWORK FOR 21ST CENTURY LEARNING

- Learning and Innovation Skills: Creativity and Innovation; Critical Thinking and Problem Solving; Communication and Collaboration
- Information, Media, and Technology Skills: Information Literacy; Media Literacy; ICT Literacy
- Life and Career Skills: Flexibility and Adaptability; Initiative and Self-Direction; Social and Cross-Cultural Skills; Productivity and Accountability; Leadership and Responsibility

Table 4.11. Key Vocabulary for Lesson 4

Key Vocabulary	Definition
absolute dating	a method of applying a chronological age to a rock layer
anchorage	"a secure fixing usually made of reinforced concrete to which the cables are fastened" (PBS 2001)
cable-stayed bridge	"a bridge in which the roadway deck is suspended from cables anchored to one or more towers" (PBS 2001)
compression	"a pressing force that squeezes material together" (PBS 2001)
crosscutting relationship	a principle that states that geological structures such as igneous intrusions and faults are younger than the rocks that they cut across
friction	the force that occurs when two objects move against each other
index fossil	a fossil that helps geologist date the rock layer (or strata) in which it was found. It can also help correlate that rock layer to other similar strata
outcrop	a rock formation that is visible on the surface; natural outcrops that occur in nature include stream beds and caverns
principle of original horizontality	a principle that states that sediment is originally deposited horizontally in an area
principle of superposition	a principle that states that given a sequence of layered rocks, the oldest rock is on the bottom and the youngest rock is on the top

Continued

Table 4.11. *(continued)*

Key Vocabulary	Definition
radiometric dating	a method of dating rocks that measures the types of radioactive isotopes present in a rock sample
relative dating	a method of dating rocks that orders rocks and geologic structures (such as folds and faults) from oldest to youngest
road cut	an exposure of the rock underlying the surface that is usually created when road construction crews blast away rock to create a road
suspension bridge	"a bridge in which the roadway deck is suspended from cables that pass over two towers; the cables are anchored in housings at either end of the bridge" (PBS 2001)
tension	a stretching force that pulls materials apart
unconformity	a point of contact between two rock layers that are very different in age; typically, an unconformity represents a gap in the geologic record

TEACHER BACKGROUND INFORMATION

Mathematics

Suspension Bridges. Many people think of suspension bridges as being a modern bridge design. While this is true of designs like the Brooklyn Bridge and the Golden Gate Bridge, suspension bridges have been used for many years. Early forms of suspension bridges typically consisted of planks of wood suspended from a rope anchored at each end of the bridge. These early bridges were mainly used for pedestrian traffic and were limited in the amount of weight they could support. However, with the development of building materials such as steel, engineers were able to design suspension bridges that could support a much larger amount of weight. These modern designs of suspension bridges were among the first bridges to have very long spans that allowed for water traffic below the bridge.

The main forces acting on the bridge are *compression, tension,* and *friction:*

- Compression can be found in the towers. The towers push up and the cable presses straight down where it wraps over the top of the tower. This causes a compression force in the tower.
- Tension is found in the main cable and the cables that run from the main cable to the bridge deck. Because the cables support the weight of the bridge, the tension force acts to stretch the cables.

- Friction is a force that we have not discussed in the other bridges. This can be found as the anchorages "resist" being pulled toward the center of the bridge. The friction force points away from the center of the bridge.

See page 172 for a drawing of a force diagram. For more information about suspension bridges, see the following websites:

- *http://science.howstuffworks.com/engineering/civil/bridge6.htm*
- *www.britannica.com/technology/suspension-bridge*

Cable-Stayed Bridges. Cable-stayed bridges are the most modern design of the bridges we have discussed so far. At first glance these bridges look very similar to suspension bridges; however, while both designs use cables to support the deck, they do so in different ways. As discussed above, the suspension bridge cables rest over top of the towers and are attached to large anchorages on either side of the bridge. In contrast, in a cable-stayed bridge, the cables are anchored directly to the towers.

The advantages of cable-stayed bridges include the following:

- They are very strong and effective for span lengths between 500 and 2,800 feet.
- Cable-stayed bridges are less expensive than suspension bridges because they require much less cable.
- They are relatively easy to build.
- Cable-stayed bridges are visually appealing.

For more information about cable-stayed bridges, see the following websites.

- *www.pbs.org/wgbh/nova/lostempires/china/meetcable.html*
- *http://science.howstuffworks.com/engineering/civil/bridge7.htm*

Science

Geologists use multiple techniques to determine the age of rocks. *Relative dating* techniques allow the geologist to sequence the rock strata in the order in which they occurred. The *principle of superposition,* put simply, states that the oldest rocks will be on the bottom and the younger rocks will be on the top in a sequence of rocks. Likewise, the *principle of crosscutting relationships* states that geologic structures such as faults and igneous intrusions are younger than the rocks in which they are present. Sometimes, there is a gap in the geologic record when deposited material has eroded. Such gaps in the geologic record are called *unconformities.*

Absolute dating techniques provide geologists with a more accurate date when the event occurred. *Radiometric dating* involves measuring the proportion of radioactive isotopes in a rock sample. This is beyond what students will do in this lesson.

For more information on geologic methods for dating rocks and fossils, please visit the following websites:

- *http://geology.utah.gov/map-pub/survey-notes/glad-you-asked/glad-you-asked-how-do-geologists-know-how-old-a-rock-is*
- *www.nature.com/scitable/knowledge/library/dating-rocks-and-fossils-using-geologic-methods-107924044*

COMMON MISCONCEPTIONS

Students will have various types of prior knowledge about the concepts introduced in this lesson. Table 4.12 outlines some common misconceptions students may have concerning these concepts. Because of the breadth of students' experiences, it is not possible to anticipate every misconception that students may bring as they approach this lesson. Incorrect or inaccurate prior understanding of concepts can influence student learning in the future, however, so it is important to be alert to misconceptions such as those presented in the table.

Table 4.12. Common Misconceptions About the Concepts in Lesson 4

Topic	Student Misconception	Explanation
Variables	There is confusion about which are the independent and dependent variables.	An independent variable is a variable that you can control (span length of a bridge). A dependent variable is a variable that you observe and measure (weight the span will hold).
Models	Models are only in 3-D.	Models in fact can be 2-D representations such as pictures, graphs, written descriptions, blueprints, equations, and other representations, as well as a 3-D model.
	A model's usefulness is based solely on the model's physical resemblance to the object being modeled.	A physical model is a smaller or larger physical copy of an object such as a bridge. The model represents a similar object in the sense that scale is an important characteristic of the model.

Continued

Table 4.12. *(continued)*

Topic	Student Misconception	Explanation
Force (a push or pull between objects)	Nonmoving objects do not exert a force (e.g., a tabletop or roadway).	Stationary objects can exert forces on other objects. For example, when you roll a toy car over a tabletop, there is a frictional force between the table and the car wheels.
	If a moving object is slowing, the force that was propelling it forward is decreasing.	When a force acts on a moving object in a direction opposite the direction of motion, the moving object will slow, even if the force that was propelling the object forward continues.
Weathering, erosion, and deposition (the breaking down of a material, the movement of the material to another area)	Water and wind cannot wear away rock and other building materials.	Water and wind can wear away rock and other building materials.
	Water freezing in cracks cannot break apart rock.	Water freezing in cracks can break apart rock.
Earth materials (types and uses of rocks)	Rocks have little practical use.	Rocks are used for a wide variety of consumer and industrial purposes. For example, rocks are used in paper production, cement, household cleaners, jewelry, pencils, chalk, glass, and building materials such as bricks and kitchen countertops.
	All rocks are the same.	There are three major types of rocks (sedimentary, igneous, and metamorphic), which have a variety of mineral compositions.

PREPARATION FOR LESSON 4

Gather and organize materials for the mathematics investigation and make copies of the student handouts attached at the end of this lesson. Reserve computers or web-enabled devices for use in science, ELA, and social studies.

LEARNING COMPONENTS

Introductory Activity/Engagement

Connection to the Challenge: Begin each day of this lesson by directing students' attention to the driving question for the module and challenge: How can we develop a decision model to help us make a recommendation to the local department of transportation on the type of bridge to build for a given location? Ask students why bridges are so important and what impact a bridge might have on a community. Hold a brief student discussion of how their learning in the previous days' lesson(s) contributed to their ability to create their plan and build their prototype. You may wish to hold a class discussion, creating a class list of key ideas on chart paper, or you may wish to have students create a notebook entry with this information.

Driving Question for Lesson 4: How are suspension bridges and cable-stayed bridges different from arch bridges and beam bridges? Do they support weight differently? Are they found in different environments?

Mathematics Class: Ask the following questions:

- What types of bridges have we explored so far?
- What are some other types of bridges we have not discussed yet?

Show students two examples of suspension bridges. Show one that is a rope bridge and another that is a large suspension bridge such as the Golden Gate Bridge. Possible images are included on page 173. Ask the following questions:

- What kind of bridges are these?
- What do they have in common?
- What is different about these two bridges?
- Are there any similarities between these bridges and arch bridges?
- What part of the bridge is supporting the weight?

Science Connection: Ask the class the following questions:

- What do road and bridge building and geology have in common?
- Why might a geologist be interested when a road construction crew uncovers the earth below?

Show students the images of road cuts on page 174. Tell the class that a road cut is a geologist's best friend. Road cuts make the geologist's job much easier since natural outcrops, such as streambeds and caverns, are not readily available at most sites in need of study. Road cuts provide a view of rocks and geological formations that would normally be below the surface. Studying these rocks and determining their ages allow geologists to learn about the history of the Earth. In this lesson, students learn about different methods that scientists use to tell how old rocks are.

ELA Connection: Continue the bridge biographies.

Social Studies Connection: Show students an image of the Waco Suspension Bridge (p. 175). Ask students if they know the life story of the Waco Suspension Bridge.

Activity/Exploration

Mathematics Class: Place the students in their project groups and give each student a Suspension Bridge Weight Test handout (p. 163). As students are working, you should move around the room to ensure students have the proper setup.

Ask the following:

- What happens to the anchorages as you add weight?
- How did your group determine the fail point?
- How did your group make sure you were being consistent from trial to trial?
- How do the towers react as you add varying amounts of weight?
- How many pennies are you adding to the anchorage cups between each trial? Should you stay consistent?
- Do you recognize a pattern between the number of pennies in each anchorage cup and the number of pennies in the weight cup?

Once each group has finished their investigation, bring the class back together to discuss their findings. Ask the following:

- What challenges did you face in this experiment?
- Did your model accurately predict the size of the anchorages used in the construction of the Golden Gate Bridge?
- Why do you think this was?
- Why do you think it took engineers longer to "perfect" this type of bridge?

After completing the discussion, have the students return to their groups and change one aspect of their setup. Possible modifications include the following:

- Height of the towers
- Distance between the desks
- Added friction to the bottom of the anchorage cups

As the students attempt to improve their setup, have them create a new model and determine whether it is a more accurate model of the relationship between the weight of the anchorages and the amount the bridge is able to support. Bring the class back together and have students share their findings.

STEM Research Notebook Prompt

Have students answer the following questions in their STEM Research Notebooks:

- *What modifications did you make to your design?*
- *What effect did the modifications have?*
- *What is another modification you could make that you believe would reduce the required weight of the anchorages?*

Science Connection: Lead students through a lesson on the geologic dating of rocks. For example, you could use the U.S. Geologic Survey's School Yard Geology lesson found at *http://education.usgs.gov/lessons/schoolyard/GeoSleuth.html,* which provides a murder mystery context for learning about some basic geologic dating methods.

ELA Connection: Connect to science, mathematics, and social studies content explored in this unit. Work with students to highlight central learning from the other content areas related to bridges. Record central concepts on a class chart, and have students address how their bridge story provides an example of the concept. Add each example to the chart underneath the central concept.

Social Studies Connection: Have students research how the building of the Waco Suspension Bridge affected people lives on both a micro (individual/family) and macro (society) level. Students will discuss how the bridge played a role in the lives of the people in the community and how its structure may have changed as a result of changes in the community.

Explanation

Mathematics Class: Give each student the Suspension Bridge Basics handout (p. 167). Have the students discuss the suspension bridge vocabulary. Building on their investigation, students should identify the role each of these terms played in their experiment and label the diagram provided on the handout.

Draw the students' attention to the forces section of the handout. It is important to note that while an arch is strong because of compression, a suspension bridge's strength comes from the tension found in the massive cables. Materials used in the construction of the cable must be able to withstand massive amounts of tension.

STEM Research Notebook Prompt

Have students journal about what they have learned about the design, function, and structure of suspension bridges.

Science Connection: Have students work in pairs to conduct internet research to find definitions of the following terms (terms and definitions should be recorded in students' STEM Research Notebooks):

- Relative dating
- Absolute dating
- Principle of superposition
- Principle of original horizontality
- Crosscutting relationship
- Unconformity
- Index fossil

Direct students to the following websites for definitions:

- *www.nps.gov/subjects/geology/geotime.htm*
- *www.nature.com/scitable/knowledge/library/dating-rocks-and-fossils-using-geologic-methods-107924044*
- *www.sciencelearn.org.nz/resources/1485-relative-dating*
- *www.sciencelearn.org.nz/resources/1486-absolute-dating*

ELA Connection: Continue the bridge biographies.

Social Studies Connection: Students write a reflection on how our experiences with bridges contribute to our cultural metaphors related to bridges.

Elaboration/Application of Knowledge

Mathematics Class: (*Note:* A warmup or mini lesson about the Pythagorean theorem may be needed at this point.) Show students another picture of the Golden Gate Bridge and a picture of the Millau Bridge (p. 176). Ask the following:

- What are some similarities between these two bridges?
- How are they different?
- Are they both suspension bridges?
- How are these bridges anchored?

Tell students that the cable-stayed bridge is the most modern design of the four bridge styles that you are looking at. The Millau Bridge is an example of a cable-stayed bridge. Both suspension and cable-stayed bridges use cables to support the weight of the bridge. However, the main difference between the two is where the cables are anchored. In a suspension bridge, the cables go over the top of the towers and are anchored to large anchorages. In a cable-stayed bridge, the cables are anchored to the towers themselves.

Give each student a Cable-Stayed Bridge Basics handout (p. 168). Give groups the opportunity to write vocabulary and draw the forces diagram. Ask the following:

- What vocabulary is similar to the suspension bridge?
- Look at the forces diagram for the suspension bridge and the cable-stayed bridge. How are they similar? How are they different?

Once students have an understanding of general vocabulary, place the students in their project groups and give them a Cable-Stayed Bridge Investigation handout (p. 169). As students are working, circulate and ask questions such as the following:

- What are your initial thoughts as you approach this problem?
- What shapes do you see that are involved in this problem?
- How can we use the fact that these are right triangles to help us solve this problem?
- I see that many of you are using the Pythagorean theorem over and over to solve this problem. What do we know about all of these triangles? Can we use the fact that these triangles are similar to help us solve the problem in a different way?

Once the students have completed their investigation, bring the class together and give groups the opportunity to share their solution methods and results.

STEM Research Notebook Prompt

Have students journal about what they have learned about the design, function, and structure of cable-stayed bridges.

Science Connection: Have students view the geologic cross section of the Grand Canyon found at *http://education.usgs.gov/images/schoolyard/GrandCanyonAge.jpg.* (You may want to print copies for student pairs.) Working in pairs, have students label the rock formations from oldest to youngest. Have student pairs compare their answers with a neighboring group.

STEM Research Notebook Prompt

Have students respond to the following prompt in their STEM Research Notebooks: Depending on the location and construction methods, geologists may be a part of the bridge design team. Explain this and consider whether this might be a consideration in your decision model.

ELA Connection: Continue the bridge biographies; to go along with their published bridge biographies, each group (or individual) should create a chart providing examples of each concept from their research.

STEM Research Notebook Prompt

Have students reflect on their chart and jot down ideas in their journal that they might need to consider in their decision model.

Social Studies Connection: Not applicable.

Evaluation/Assessment

Students may be assessed on the following performance tasks and other measures listed.

Performance Tasks

- End of Lesson Assessment: Have students complete the Bridges: Compare and Contrast Matrix handout.

Other Measures

- STEM Research Notebook Entries: You should regularly read and respond to students in their STEM Research Notebooks. Your response should not indicate whether students' entries are right or wrong. Instead, include comments or questions that will push and stretch students' thinking and can aid students in moving toward development of their decision models.
- Learning Activity Responses: Suspension Bridge Weight Test, Suspension Bridge Basics, Cable-Stayed Bridge Basics, and Cable-Stayed Bridge Investigation handouts
- Quick Check (p. 177)

INTERNET RESOURCES

Information about suspension bridges

- *http://science.howstuffworks.com/engineering/civil/bridge6.htm*
- *www.britannica.com/technology/suspension-bridge*

Information about cable-stayed bridges

- *www.pbs.org/wgbh/nova/lostempires/china/meetcable.html*
- *http://science.howstuffworks.com/engineering/civil/bridge7.htm*

Information about geologic methods for dating rocks and fossils

- *http://geology.utah.gov/map-pub/survey-notes/glad-you-asked/glad-you-asked-how-do-geologists-know-how-old-a-rock-is*
- *www.nature.com/scitable/knowledge/library/dating-rocks-and-fossils-using-geologic-methods-107924044*

Resources for definitions of rock dating terms

- *www.nps.gov/subjects/geology/geotime.htm*
- *www.nature.com/scitable/knowledge/library/dating-rocks-and-fossils-using-geologic-methods-107924044*
- *www.sciencelearn.org.nz/resources/1485-relative-dating*
- *www.sciencelearn.org.nz/resources/1486-absolute-dating*

U.S. Geologic Survey's School Yard Geology lesson

- *http://education.usgs.gov/lessons/schoolyard/GeoSleuth.html*

Geologic cross section of the Grand Canyon

- *http://education.usgs.gov/images/schoolyard/GrandCanyonAge.jpg*

Name: ______________________________

STUDENT HANDOUT, PAGE 1

SUSPENSION BRIDGE WEIGHT TEST

Many view suspension bridges as being relatively modern; however, primitive designs of suspension bridges have been used for a very long time. Using rope bridges, early designers could span relatively long distances without needing supports in the middle of the bridge. These bridges were fairly strong but were almost exclusively used to allow individuals to walk from one side to the other. However, with the advancement in building materials, engineers have been able to design and build suspension bridges that can span very large distances and support enormous amounts of weight. Your task is to explore the relationship between the maximum weight that is able to be supported using anchorages of varying weights.

MATERIALS

- Clear tape
- 1 paper clip
- Drinking straws (4 per group)
- String (approximately 2 feet per group)
- 2 oz. plastic cups (3 per group)
- Pennies (100 per group; similar weights may be used if pennies are not an option)
- Safety glasses or safety goggles

SAFETY NOTES

1. All laboratory occupants must wear safety glasses or goggles during all phases of this inquiry activity.
2. Use caution when working with sharps (e.g., scissors) to avoid cutting or puncturing skin.
3. Make sure all materials are put away after completing the activity.
4. Wash hands with soap and water after completing this activity.

SETUP

1. Use scissors to cut 1 inch off each of the four drinking straws.
2. Use the clear tape to attach two straws together and place one of the 1-inch cut pieces horizontally across the top of the straws (see diagram above).

Name: ____

STUDENT HANDOUT, PAGE 2

SUSPENSION BRIDGE WEIGHT TEST

3. Attach the ends of the string to the rim of the 2 oz. plastic cups.
4. Place two desks 6 inches apart and tape the "piers" made in step 2 to the edge of the desk. Place the string over the piers as shown in the diagram below.
5. Use a small piece of string and the paper clip to attach the third cup to the middle of the string (main cable).

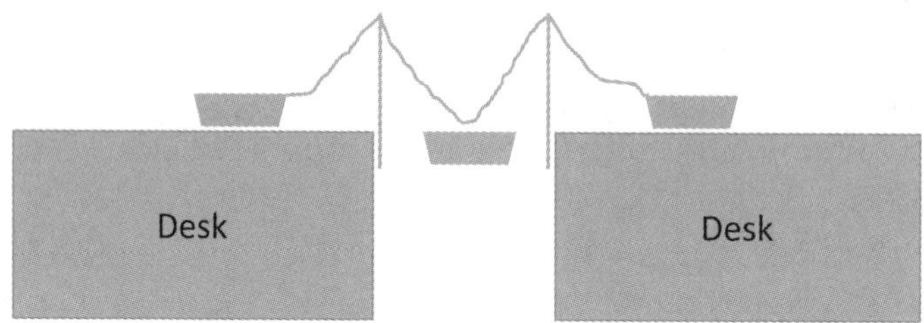

6. Place the plastic cups on the desk so that the cup in the middle is even with the tops of the desks. These cups act as your anchorages (referred to as *anchorage cups* throughout the activity). Place a mark on each desk so you are able to return the cups to their original position.
7. Place equal amounts of pennies in the two cups and then add weight to the center cup until the bridge fails. DO NOT drop the pennies in the cup. Have one person support the cup while another person gently places the pennies in the cup.
8. Explore the setup and decide as a group what qualifies as a bridge collapse before you officially start collecting data.

Number of Pennies in Each Anchorage Cup	Number of Pennies *Before* Collapse	Observations

Name: ___________________________________

STUDENT HANDOUT, PAGE 3

SUSPENSION BRIDGE WEIGHT TEST

INVESTIGATION QUESTIONS

1. What are the variables involved in this problem?

2. How can you organize the data to give a clearer picture of what is happening?

3. Describe the relationship between your variables.

4. How could you construct a model (equation) that relates the variables?

5. How many pennies would you need to place in the anchorage cups to support the weight of 200 pennies?

Name: ______________________________

STUDENT HANDOUT, PAGE 4

SUSPENSION BRIDGE WEIGHT TEST

6. The deck of the Golden Gate Bridge weighs approximately 420,000 tons. According to your model, how much would the anchorages need to weigh to support that much weight?

7. The anchorages for the Golden Gate Bridge weigh approximately 60,000 tons. How does this compare with your answer to Question 6? How can you account for this difference?

8. When the bridge fails, what happens to the anchorage cups? How could you help these anchorages stay in place WITHOUT adding more weight?

9. What would your observations about your bridge failure and the anchorage cups mean for city planners as they try to determine whether a suspension bridge is a suitable design for a particular site?

10. What part of the bridge is not a part of the experiment? What effect does that have on the experiment?

Name: ___________________________________

STUDENT HANDOUT

SUSPENSION BRIDGE BASICS

Write the definition of each of the following terms. Be sure to label your picture.

Tower:

Anchorage:

Suspension cable:

Deck:

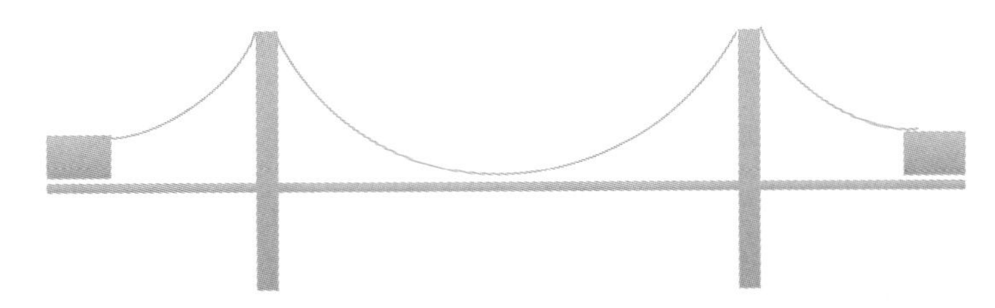

Explain how the following forces act on a suspension bridge. Be sure to draw these forces on the diagram. Remember, for the bridge not to move, all forces must cancel out.

Compression:

Tension:

Friction:

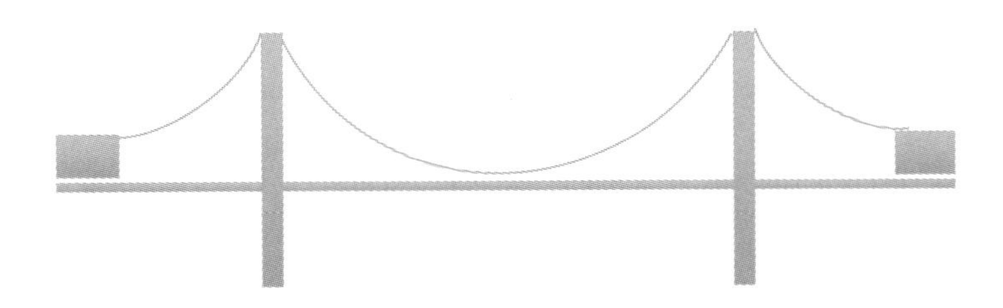

Name: ______________________________

STUDENT HANDOUT

CABLE-STAYED BRIDGE BASICS

Write the definition of each of the following terms. Be sure to label your picture.

Tower:

Cable (Stays):

Deck:

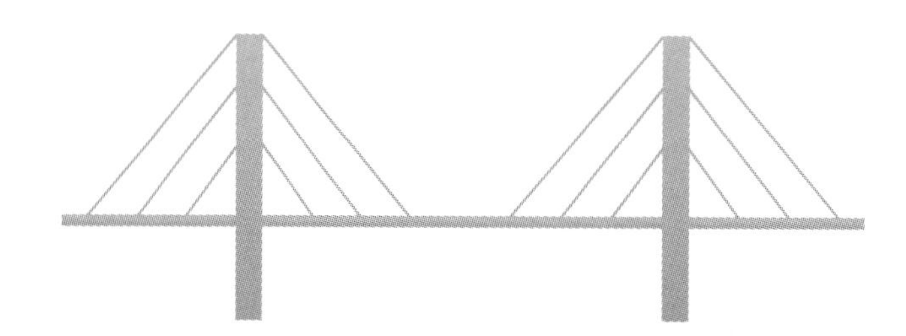

Explain how the following forces act on a cable-stayed bridge. Be sure to draw these forces on the diagram. Remember, for the bridge not to move, all forces must cancel out.

Compression:

Tension:

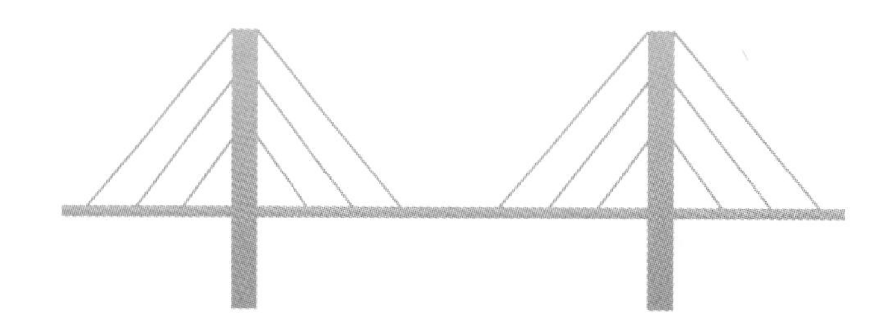

Name: ______________________________

STUDENT HANDOUT, PAGE 1

CABLE-STAYED BRIDGE INVESTIGATION

The cable-stayed bridge is a modern design that is architecturally beautiful and very strong. However, it is also one of the most expensive bridge designs. For this reason, the city's civil engineers want to make sure they have an accurate estimate of the amount of material that is needed to construct the cables. The picture below is a simple diagram of a cable-stayed bridge. Use it to help you investigate the problem that follows.

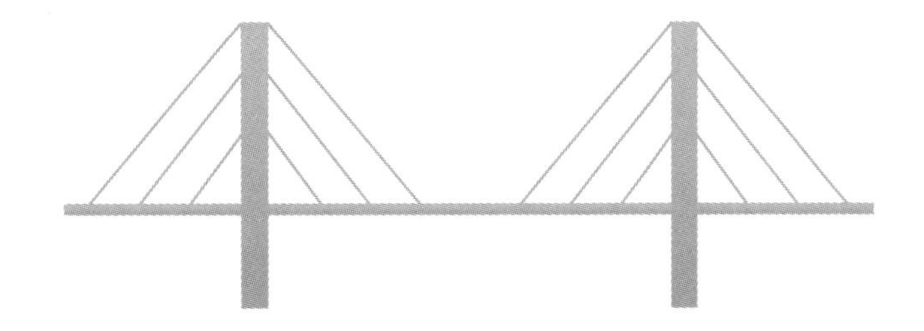

PROBLEM

Civil engineers have determined that they will need 13 cables on each side of two different towers to support the needed weight and give the appropriate span length. The first cable will attach to the tower 15 feet above the deck and will attach to the deck 25 feet from the base of the tower. Additional cables will attach to the tower every 15 feet and will be parallel to the cable below it. Determine how many feet of cable will be needed to construct a bridge with two towers.

INVESTIGATION QUESTIONS

1. What are your initial thoughts as you begin this problem?

Name: ______________________________

STUDENT HANDOUT, PAGE 2

CABLE-STAYED BRIDGE INVESTIGATION

2. Draw a picture that represents the problem.

3. How many feet of cable would be needed to construct two complete towers with 13 sets of cables each? Please describe how your group solved the problem.

4. Using these two towers, how long of a span do you think can be created? Think back to what you learned about beam bridges.

Name: ______________________________

STUDENT HANDOUT

BRIDGES: COMPARE AND CONTRAST MATRIX

Attribute	Type of Bridge			
	Beam	**Arch**	**Suspension**	**Cable-Stayed**
Design				
Structure				
Function				
Strength				
Span Length				

FORCE DIAGRAM

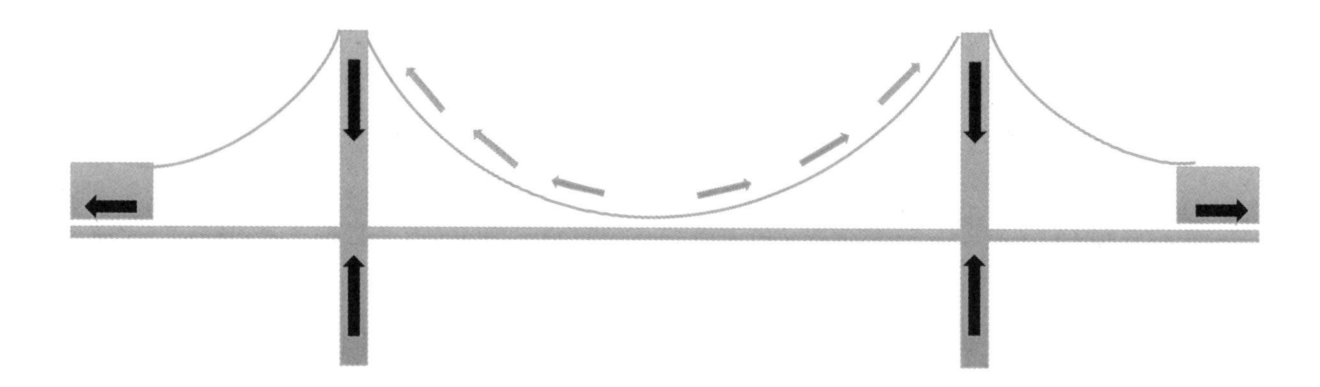

EXAMPLES OF SUSPENSION BRIDGES

Rope Bridge

Golden Gate Bridge

IMAGES OF ROAD CUTS

WACO SUSPENSION BRIDGE

COMPARING THE GOLDEN GATE BRIDGE AND THE MILLAU BRIDGE

Golden Gate Bridge

Millau Bridge

Name: ___________________________________

STUDENT HANDOUT

QUICK CHECK

Some engineers believe that the design we used is not the strongest design for a cable-stayed bridge. Instead they argue that the following design is stronger:

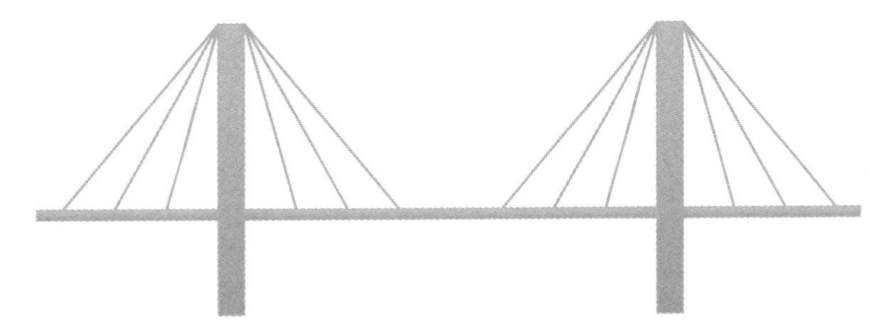

Assume that all the cables attach to the towers at a height of 195 feet above the deck. To make sure the bridge is strong, the 13 cables still need to attach to the deck every 25 feet from the tower.

1. Draw a diagram to represent the problem.

2. Does this design require more or less cable then the design we discussed in class?

3. Which design do YOU think would be the strongest? It might help to think about the investigation we did with suspension bridges.

Lesson Plan 5: Economics and Bridges

Bridges can have a major impact on the people and communities in which they are found. In this lesson, students explore some of the economic impacts bridges have had in our society. In mathematics, students will be presented with four plans that describe the initial and future costs of building a bridge. Students will be asked to analyze these options and determine which plan is the best choice for various periods of time.

ESSENTIAL QUESTIONS

- How do we determine which cost breakdown is the best choice when initial and maintenance costs vary?
- What was the social and economic impact of the Works Progress Administration (WPA)?

ESTABLISHED GOALS AND OBJECTIVES

At the conclusion of this lesson, students will be able to do the following:

- Write an equation for total cost given initial cost and cost of yearly maintenance.
- Graph cost equation on a coordinate plane and describe what the y-intercept and slope mean.
- Compare cost functions to determine which is the cheapest for a given time period.
- Explore the social and economic impact of the WPA.

TIME REQUIRED

- 2 days (approximately 45 minutes each day; see Tables 3.8–3.9, p. 40)

MATERIALS

- STEM Research Notebooks
- Handouts (attached at the end of this lesson)
- Graph paper

CONTENT STANDARDS AND KEY VOCABULARY

Table 4.13 lists the content standards from the *NGSS, CCSS,* and the Framework for 21st Century Learning that this lesson addresses, and Table 4.14 (p. 183) presents the key vocabulary. Vocabulary terms are provided for both teacher and student use. Teachers may choose to introduce some or all of the terms to students.

Table 4.13. Content Standards Addressed in STEM Road Map Module Lesson 5

NEXT GENERATION SCIENCE STANDARDS

PERFORMANCE EXPECTATIONS

- MS-ESS3-1. Construct a scientific explanation based on evidence for how the uneven distributions of Earth's mineral, energy, and groundwater resources are the result of past and current geoscience processes.

SCIENCE AND ENGINEERING PRACTICES

Developing and Using Models

Modeling in 6–8 builds on K–5 experiences and progresses to developing, using, and revising models to describe, test, and predict more abstract phenomena and design systems.

- Develop and use a model to describe phenomena.
- Develop a model to generate data to test ideas about designed systems, including those representing inputs and outputs.

Analyzing and Interpreting Data

Analyzing data in 6–8 builds on K–5 experiences and progresses to extending quantitative analysis to investigations, distinguishing between correlation and causation, and basic statistical techniques of data and error analysis.

- Analyze and interpret data to provide evidence for phenomena.

Constructing Explanations and Designing Solutions

Constructing explanations and designing solutions in 6–8 builds on K–5 experiences and progresses to include constructing explanations and designing solutions supported by multiple sources of evidence consistent with scientific ideas, principles, and theories.

- Construct a scientific explanation based on valid and reliable evidence obtained from sources (including the students' own experiments) and the assumption that theories and laws that describe the natural world operate today as they did in the past and will continue to do so in the future.

Scientific Knowledge Is Open to Revision in Light of New Evidence (Nature of Science practice)

- Science findings are frequently revised and/or reinterpreted based on new evidence.

DISCIPLINARY CORE IDEAS

ESS2.A: Earth's Materials and Systems

- All Earth processes are the result of energy flowing and matter cycling within and among the planet's systems. This energy is derived from the sun and Earth's hot interior. The energy that flows and matter that cycles produce chemical and physical changes in Earth's materials and living organisms.

Continued

Table 4.13. *(continued)*

DISCIPLINARY CORE IDEAS *(continued)*

ESS3.A: Natural Resources

- Humans depend on Earth's land, ocean, atmosphere, and biosphere for many different resources. Minerals, fresh water, and biosphere resources are limited, and many are not renewable or replaceable over human lifetimes. These resources are distributed unevenly around the planet as a result of past geologic processes.

CROSSCUTTING CONCEPTS

Stability and Change

- Explanations of stability and change in natural or designed systems can be constructed by examining the changes over time and processes at different scales, including the atomic scale.

Patterns

- Patterns in rates of change and other numerical relationships can provide information about natural systems.

Cause and Effect

- Cause and effect relationships may be used to predict phenomena in natural or designed systems.

Influence of Science, Engineering, and Technology on Society and the Natural World

- All human activity draws on natural resources and has both short- and long-term consequences, positive as well as negative, for the health of people and the natural environment.

COMMON CORE STATE STANDARDS FOR MATHEMATICS

MATHEMATICAL PRACTICES

- MP1. Make sense of problems and persevere in solving them.
- MP2. Reason abstractly and quantitatively.
- MP3. Construct viable arguments and critique the reasoning of others.
- MP6. Attend to precision.

MATHEMATICAL CONTENT

- 8.EE.B.5. Graph proportional relationships, interpreting the unit rate as the slope of the graph. Compare two different proportional relationships represented in different ways. For example, compare a distance-time graph to a distance-time equation to determine which of two moving objects has greater speed.
- 8.EE.C.7.B. Solve linear equations with rational number coefficients, including equations whose solutions require expanding expressions using the distributive property and collecting like terms.

Continued

Table 4.13. (*continued*)

MATHEMATICAL CONTENT (*continued*)

- 8.F.B.5. Describe qualitatively the functional relationship between two quantities by analyzing a graph (e.g., where the function is increasing or decreasing, linear or nonlinear). Sketch a graph that exhibits the qualitative features of a function that has been described verbally.

COMMON CORE STATE STANDARDS FOR ENGLISH LANGUAGE ARTS

(Note: These standards arch over the entire bridge biography project and include interactions within the writing workshop.)

READING STANDARD

- RI.8.7. Evaluate the advantages and disadvantages of using different mediums (e.g., print or digital text, video, multimedia) to present a particular topic or idea.

WRITING STANDARDS

- W.8.1.E. Provide a concluding statement or section that follows from and supports the argument presented.
- W.8.2. Write informative/explanatory texts to examine a topic and convey ideas, concepts, and information through the selection, organization, and analysis of relevant content.
- W.8.2.A. Introduce a topic clearly, previewing what is to follow; organize ideas, concepts, and information into broader categories; include formatting (e.g., headings), graphics (e.g., charts, tables), and multimedia when useful to aiding comprehension.
- W.8.2.B. Develop the topic with relevant, well-chosen facts, definitions, concrete details, quotations, or other information and examples.
- W.8.2.C. Use appropriate and varied transitions to create cohesion and clarify the relationships among ideas and concepts.
- W.8.2.D. Use precise language and domain-specific vocabulary to inform about or explain the topic.
- W.8.2.E. Establish and maintain a formal style.
- W.8.2.F. Provide a concluding statement or section that follows from and supports the information or explanation presented.
- W.8.3. Write narratives to develop real or imagined experiences or events using effective technique, relevant descriptive details, and well-structured event sequences.
- W.8.3.A. Engage and orient the reader by establishing a context and point of view and introducing a narrator and/or characters; organize an event sequence that unfolds naturally and logically.
- W.8.3.D. Use precise words and phrases, relevant descriptive details, and sensory language to capture the action and convey experiences and events.

Continued

Table 4.13. *(continued)*

WRITING STANDARDS (*continued*)

- W.8.6. Use technology, including the internet, to produce and publish writing and present the relationships between information and ideas efficiently as well as to interact and collaborate with others.
- W.8.8. Gather relevant information from multiple print and digital sources, using search terms effectively; assess the credibility and accuracy of each source; and quote or paraphrase the data and conclusions of others while avoiding plagiarism and following a standard format for citation.

SPEAKING AND LISTENING STANDARDS

- SL.8.1. Engage effectively in a range of collaborative discussions (one-on-one, in groups, and teacher-led) with diverse partners on grade 8 topics, texts, and issues, building on others' ideas and expressing their own clearly.
- SL.8.1.A. Come to discussions prepared, having read or researched material under study; explicitly draw on that preparation by referring to evidence on the topic, text, or issue to probe and reflect on ideas under discussion.
- SL.8.1.B. Follow rules for collegial discussions and decision making, track progress toward specific goals and deadlines, and define individual roles as needed.
- SL.8.1.C. Pose questions that connect the ideas of several speakers and respond to others' questions and comments with relevant evidence, observations, and ideas.
- SL.8.1.D. Acknowledge new information expressed by others, and, when warranted, qualify or justify their own views in light of the evidence presented.
- SL.8.4. Present claims and findings, emphasizing salient points in a focused, coherent manner with relevant evidence, sound valid reasoning, and well-chosen details; use appropriate eye contact, adequate volume, and clear pronunciation.
- SL.8.5. Integrate multimedia and visual displays into presentations to clarify information, strengthen claims and evidence, and add interest.
- SL.8.6. Adapt speech to a variety of contexts and tasks, demonstrating command of formal English when indicated or appropriate.

FRAMEWORK FOR 21ST CENTURY LEARNING

- Interdisciplinary Themes: Financial, Economic, Business, and Entrepreneurial Literacy; Civic Literacy
- Learning and Innovation Skills: Creativity and Innovation; Critical Thinking and Problem Solving; Communication and Collaboration
- Information, Media, and Technology Skills: Information Literacy; Media Literacy; ICT Literacy
- Life and Career Skills: Flexibility and Adaptability; Initiative and Self-Direction; Social and Cross-Cultural Skills; Productivity and Accountability; Leadership and Responsibility

Table 4.14. Key Vocabulary for Lesson 5

Key Vocabulary	Definition
bid	an estimate of how much it will cost to complete a service or project
cost	the amount of money that will be needed to purchase or pay for something
cross section	a graphical representation of a vertical slice through rock layers (much like a slice of a layer cake)
maintenance cost	the amount of money that is required to maintain an item's safety or quality
piecewise function	a function that is composed of multiple subfunctions whose domains do not overlap

TEACHER BACKGROUND INFORMATION

Mathematics

In this lesson, students are presented with four descriptions of initial and future costs of a bridge. During the investigation, students will be asked to discuss these four plans and determine the pros and cons of each. Once students have thoroughly explored the four options, they will represent the cost of each plan with an equation as well as on a graph. This will help them determine which is the best choice for a given time period. Before beginning this lesson, note that plans C and D are piecewise defined functions. While students may not have had experience with piecewise functions at this point, they should be able to reason through drawing the graph of these functions. However, as students work to write their models that give the total cost, you may need to provide support in writing their equations in the correct notation.

Social Studies

From 1929 to 1939, the United States fell into the Great Depression. This was the United States' largest economic downturn. Banks failed, the stock market crashed, and millions of Americans were unemployed. President Franklin D. Roosevelt developed the WPA to help boost the economy and put Americans back to work. The WPA employed 8.5 million Americans from 1935 to 1943, which in turn helped the economy grow. For more information on the Great Depression and WPA, visit the following websites:

- *www.history.com/topics/great-depression*
- *www.pbs.org/wgbh/americanexperience/features/surviving-the-dust-bowl-works-progress-administration-wpa*
- *www.history.com/this-day-in-history/fdr-creates-the-wpa*
- *http://rooseveltinstitute.org/works-progress-administration*

COMMON MISCONCEPTIONS

Students will have various types of prior knowledge about the concepts introduced in this lesson. Table 4.15 outlines some common misconceptions students may have concerning these concepts. Because of the breadth of students' experiences, it is not possible to anticipate every misconception that students may bring as they approach this lesson. Incorrect or inaccurate prior understanding of concepts can influence student learning in the future, however, so it is important to be alert to misconceptions such as those presented in the table.

Table 4.15. Common Misconceptions About the Concepts in Lesson 5

Topic	Student Misconception	Explanation
Variables	There is confusion about which are the independent and dependent variables.	An independent variable is a variable that you can control (span length of a bridge). A dependent variable is a variable that you observe and measure (weight the span will hold).
Models	Models are only in 3-D.	Models in fact can be 2-D representations such as pictures, graphs, written descriptions, blueprints, equations, and other representations, as well as a 3-D model.

PREPARATION FOR LESSON 5

Gather materials for the mathematics investigation and make copies of student handouts attached at the end of this lesson.

LEARNING COMPONENTS

Introductory Activity/Engagement

Connection to the Challenge: Begin each day of this lesson by directing students' attention to the driving question for the module and challenge: How can we develop a decision model to help us make a recommendation to the local department of transportation on the type of bridge to build for a given location? Ask students why bridges are so important and what impact a bridge might have on a community. Hold a brief student discussion of how their learning in the previous days' lesson(s) contributed to their ability to create their plan and build their prototype. You may wish to hold a class discussion, creating a class list of key ideas on chart paper, or you may wish to have students create a notebook entry with this information.

Driving Questions for Lesson 5: What are the economic factors involved in building a bridge? What resources are needed?

Mathematics Class: As the students come in the door, have the following information on the board: *The Golden Gate Bridge is one of the most famous bridges in the world. Construction of the Golden Gate Bridge began January 5, 1933. Construction lasted approximately four years and the total cost of the project exceeded $35 million.*

STEM Research Notebook Prompt

Have students answer the following questions in their STEM Research Notebooks individually. Once students have reflected individually, encourage them to discuss their ideas with their groups.

- *If the Golden Gate Bridge were started now, how long do you think it would take to build?*
- *Do you think the project would cost more or less than it did in the 1930s?*
- *How much more or less?*
- *What factors do you think caused this difference in price?*
- *List as many variables you can come up with that add to the cost of building a bridge.*

Science Connection: In this lesson, students consider how humans use Earth materials and how the geological composition of the Earth affects the placement and construction of roads and bridges.

STEM Research Notebook Prompt

Have students respond to the following questions in their STEM Research Notebooks:

- *Why might a civil engineer work with a geologist to design a bridge?*
- *How many pounds of Earth materials do you think you use each year?*

ELA Connection: Bridge biographies should be nearing publication stage. Discuss how students will share their work with a broader audience. Options include a browsing room that peers, parents, and teachers may visit; presentations to small groups (e.g., other classes, community groups); and in-class presentations. Decide on how to share and prepare accordingly. Schedule the remaining ELA sessions according to need as students finalize and share their biographies.

Social Studies Connection: Show students a WPA stamp. (Images of WPA stamps can be found at *www.milfordhistory.org/wpa_sidewalks_of_milford.html* and *www.idahostatesman.com/2013/02/28/2469898/sidewalk-stamps.html*.) Ask the following questions:

- Have you ever seen stamps like these around our community?
- Do you know what they mean?

Have students discuss what they believe the significance of the stamp is.

Activity/Exploration

Mathematics Class: Pose the following scenario: For many building projects, companies present a bid to the city in hopes of being awarded the job. By making a bid, companies are promising that they will build the desired bridge for a stated amount of money. Because companies are competing for business, they want to place the lowest bid possible so they will be given the contract.

- What will happen if their bid is too high?
- What will happen if their bid is too low?

For these reasons, companies want to make sure that their estimates are as accurate as possible.

Give each student a Cost of Bridges Investigation—Picking the Best Option handout (p. 190) and give them an opportunity to read and explore the four plans. Inform students that at this point they are not to "solve" the problem. Instead, students should read each plan carefully and explore pros and cons of each plan. As the students are exploring the four plans, you should circulate around the room and ask the following questions:

- Which plan do you think will cost the most? Which plan do you think will cost the least?
- After the initial cost, which plan costs the most to maintain? Why do you think that is?

Science Connection: Have students work in pairs to conduct internet research to find answers to the following questions and record their answers in their STEM Research Notebooks:

- Why might a civil engineer work with a geologist to design a bridge?
- How many pounds of Earth materials do you use each year?
- What Earth materials are used in manufacturing a car?
- What kind of Earth materials are used in healthcare?

The following websites may be useful to students as they search for answers to these questions:

- *www.americangeosciences.org/critical-issues/faq/how-many-pounds-minerals-are-required-average-person-year*
- *www.oum.ox.ac.uk/thezone/minerals/index.htm*

ELA Connection: Not applicable.

Social Studies Connection: Have students research the WPA:

- Why was it started?
- What was its purpose?
- What was its impact on your community, state, and nation?

Be sure to focus students' attention on projects in their own community and state.

Explanation

Mathematics Class: Bring students together for a short debrief. Have students explain each of the four plans.

STEM Research Notebook Prompt

Have the students answer the following prompt: *Briefly describe each of the four plans and your initial thoughts of which may be the best.*

Science Connection: Students consider where resources come from by explaining their source to a fictional fifth-grade audience in a STEM Research Notebook entry.

STEM Research Notebook Prompt

Write an explanation to a fifth grader to explain where "everything comes from." Where do we get our resources?

ELA Connection: Not applicable.

Social Studies Connection: Be sure to highlight that the WPA was one of President Franklin Delano Roosevelt's New Deal actions to help America come out of the Great Depression.

Elaboration/Application of Knowledge

Mathematics Class: Place the students back in their project groups. Tell students, "Yesterday we explored four plans to build a local bridge. Today we are going to take a closer look at each of those plans and attempt to determine which plan we feel is the best choice." Give each student the Cost of Bridges Investigation: Day 2 handout (p. 191). Ask the following questions as the students are working on their investigation:

- Where is the initial cost displayed on the graph? What do we call this?
- Where can you find the yearly maintenance cost on the graph? What do we call this number?

Once students have completed their investigation, have them share their answers to Questions 4 and 5.

STEM Research Notebook Prompt

Have students respond to the following prompt in their STEM Research Notebooks: *Briefly describe the factors that your group considered when determining the "best" plan. Was cost your only factor?*

Science Connection: Ask the class the following: Where do you find Earth materials in your everyday life? For homework, have students make a list of all of the products that they use in a typical day. The students must identify if the product is mined or grown. Some products may be both (e.g., vegetables—a grown product—may be packaged in plastic made from petroleum—a mined product). When students return to class with their lists, have them compare their findings in small groups and then work as a class to create a master list.

STEM Research Notebook Prompt

Have students journal about the impact of geology on the design, function, and structure of bridges and how considerations of the geology of a location might play a factor in developing their decision model.

ELA Connection: Share the poem "West of Here" by William Stafford. Talk together about the absence of the bridge and what it implies about the importance of bridges—actual and metaphorical—in our lives and culture.

Social Studies Connection: Not applicable.

Evaluation/Assessment

Students may be assessed on the following performance task and other measures listed.

Performance Task

- Cost of Bridges Investigation

Other Measures

- STEM Research Notebook Entries. You should regularly read and respond to students in their STEM Research Notebooks. Your response should not indicate whether students' entries are right or wrong. Instead, include comments or questions that will push and stretch students' thinking and can aid students in moving toward development of their decision models.

- Responses to investigations/activities
- Responses to investigation sheets

INTERNET RESOURCES

Information about the Great Depression and WPA

- *www.history.com/topics/great-depression*
- *www.pbs.org/wgbh/americanexperience/features/surviving-the-dust-bowl-works-progress-administration-wpa*
- *www.history.com/this-day-in-history/fdr-creates-the-wpa*
- *http://rooseveltinstitute.org/works-progress-administration*

Images of WPA stamps

- *www.milfordhistory.org/wpa_sidewalks_of_milford.html*
- *www.idahostatesman.com/2013/02/28/2469898/sidewalk-stamps.html*

Information about use of Earth materials

- *www.americangeosciences.org/critical-issues/faq/how-many-pounds-minerals-are-required-average-person-year*
- *www.oum.ox.ac.uk/thezone/minerals/index.htm*

Name: ___________________________________

STUDENT HANDOUT

COST OF BRIDGES INVESTIGATION— PICKING THE BEST OPTION

As we have discussed in class, bridges are extremely expensive and can take years to build. Before construction begins, it important to take into account both construction costs as well as maintenance costs. Below are four plans for construction and maintenance cost for a bridge that will be constructed locally. Any of the four choices will produce a bridge that meets the needs of the community. With your group, you will explore the four plans and make a recommendation for which is the best financial option.

PLAN A

Initial cost and design of the bridge will be approximately $90 million. The materials in this plan are widely used and easy to come by, bringing down the cost of materials. Engineers estimate that on average it will cost $1.5 million each year to maintain the bridge.

PLAN B

Initial cost and design of the bridge will be approximately $120 million. The materials used in Plan B are engineered to last longer. This means that while the initial cost may be more, engineers project that the average yearly cost of maintenance will only be $500,000.

PLAN C

Initial cost and design of the bridge will be approximately $100 million. This plan is not expected to require yearly maintenance. Instead, every 10 years, $8 million will need to be spent for large-scale maintenance and upkeep.

PLAN D

Initial cost and design of the bridge will be $70 million. The materials used in this bridge are safe but are much less expensive than those of the other designs. Because of this, engineers project that during the first 10 years the average yearly cost of maintenance will be $3 million per year. After that, they believe it will cost $5 million per year for the remaining life of the bridge.

Name: ____

STUDENT HANDOUT, PAGE 1

COST OF BRIDGES INVESTIGATION: DAY 2

Yesterday you explored the four plans and discussed pros and cons of each. Now you and your group are going to determine which plan is the best option and explain how you came to that conclusion.

1. Write an equation that describes the total cost of each of the four plans for t number of years.

2. Graph each of the four equations. How does having these four equations on one graph help you gain a better understanding of the problem?

Name: ____

STUDENT HANDOUT, PAGE 2

COST OF BRIDGES INVESTIGATION: DAY 2

3. Using your equations from Question 1, and your graph from Question 2, determine which plan is the best option for the first 10 years. The first 25 years. Beyond 40 years.

4. Using the information you found for Questions 1–3, determine in your group which plan you believe is the best option. Describe in complete sentences why you believe this to be true.

5. Take a look at the other three plans. Can you think of a scenario in which these other plans might be the best option?

Lesson Plan 6: Putting It All Together—Decision Models

Students will use the knowledge they have gained through their research, investigations, and classroom discussions to compare and contrast strength, span length, structure, and function of the four bridge designs they have explored in this module. By the end of this lesson, students will have completed their decision model, developed a presentation, and presented to their peers, members of the local department of transportation, and other members of the community.

ESSENTIAL QUESTIONS

- How can we help the local department of transportation choose a bridge design for a given scenario?
- What was the social and economic impact of the Works Progress Administration (WPA)?
- How can decision models help people make a better decision?

ESTABLISHED GOALS AND OBJECTIVES

At the conclusion of this lesson, students will be able to do the following:

- Develop and write a decision model to help the local department of transportation select a bridge design for a given scenario.
- Defend a position on whether another WPA should be established.

TIME REQUIRED

- 9 days (approximately 45 minutes each day; see Tables 3.9–3.10, p. 40)

MATERIALS

- STEM Research Notebooks
- Handouts (attached at the end of this lesson)
- Computers or other web-enabled devices for research and presentation development

CONTENT STANDARDS AND KEY VOCABULARY

Table 4.16 lists the content standards from the *NGSS, CCSS,* and the Framework for 21st Century Learning that this lesson addresses, and Table 4.17 presents the key vocabulary. Vocabulary terms are provided for both teacher and student use. Teachers may choose to introduce some or all of the terms to students.

Table 4.16. Content Standards Addressed in STEM Road Map Module Lesson 6

NEXT GENERATION SCIENCE STANDARDS

Not applicable

COMMON CORE STATE STANDARDS FOR MATHEMATICS

MATHEMATICAL PRACTICES

- MP1. Make sense of problems and persevere in solving them.
- MP3. Construct viable arguments and critique the reasoning of others.
- MP4. Model with mathematics.
- MP8. Look for and express regularity in repeated reasoning.

COMMON CORE STATE STANDARDS FOR ENGLISH LANGUAGE ARTS

WRITING STANDARDS

- W.8.2.A. Introduce a topic clearly, previewing what is to follow; organize ideas, concepts, and information into broader categories; include formatting (e.g., headings), graphics (e.g., charts, tables), and multimedia when useful to aiding comprehension.
- W.8.3. Write narratives to develop real or imagined experiences or events using effective technique, relevant descriptive details, and well-structured event sequences.

SPEAKING AND LISTENING STANDARD

- SL.8.3. Delineate a speaker's argument and specific claims, evaluating the soundness of the reasoning and relevance and sufficiency of the evidence and identifying when irrelevant evidence is introduced.

FRAMEWORK FOR 21ST CENTURY LEARNING

- Interdisciplinary Themes: Financial, Economic, Business, and Entrepreneurial Literacy; Civic Literacy
- Learning and Innovation Skills: Creativity and Innovation; Critical Thinking and Problem Solving; Communication and Collaboration
- Information, Media, and Technology Skills: Information Literacy; Media Literacy; ICT Literacy
- Life and Career Skills: Flexibility and Adaptability; Initiative and Self-Direction; Social and Cross-Cultural Skills; Productivity and Accountability; Leadership and Responsibility

Table 4.17. Key Vocabulary for Lesson 6

Key Vocabulary	Definition
debate	a structured discussion in which views are expressed by participants with opposing views
decision model	a model used by business professionals to aid them in making accurate predictions and decisions while avoiding bias

TEACHER BACKGROUND INFORMATION

Mathematics

Decision models are created to aid professionals in making good decisions. They combine large amounts of data and mathematical algorithms to develop a model that aids the professional in making unbiased decisions. For a more in-depth look at decision models, visit the following websites:

- *www.web-books.com/eLibrary/ON/B0/B58/070MB58.html*
- *www.mckinsey.com/insights/strategy/the_benefits_and_limits_of_decision_models*
- *www.business-analysis-made-easy.com/Decision-Making-Models.html*
- *www.ncbi.nlm.nih.gov/pmc/articles/PMC2245736*
- *www.mindtools.com/pages/article/newTED_91.htm*

Conducting an internet search for images of decision models (e.g., examples of decision-making trees) will provide you with a variety of examples that you could use with your students to illustrate visuals for their decision models.

Social Studies

Students will be debating a fictional proposal to launch a new a program similar to the WPA program of the 1940s. The class will form two teams for the debate, one for and one against the new program. You should be prepared to offer students a structure for the debate. For example, you may wish to use the following as general guidelines for the debate:

Each team should agree on two or three major points to make their argument. Have the "pro" group speak first, allowing 5 minutes for the argument. After they present their argument, allow the "con" group 3 minutes for a response, and then allow the "pro" group an additional 2 minutes for their rebuttal of the "con" group's response. Allow questions from the audience or judging panel (about 3 minutes), and then have the "con" group present their argument (5 minutes), followed by a 3-minute response from the "pro" group, and a 2-minute rebuttal from the "con" group. Next, allow the audience or judging panel

to ask questions for about 3 minutes. After this, give the "pro" group 2 minutes for a concluding statement and the "con" group 2 minutes for a concluding statement.

COMMON MISCONCEPTIONS

Students will have various types of prior knowledge about the concepts introduced in this lesson. Table 4.18 outlines some common misconceptions students may have concerning these concepts. Because of the breadth of students' experiences, it is not possible to anticipate every misconception that students may bring as they approach this lesson. Incorrect or inaccurate prior understanding of concepts can influence student learning in the future, however, so it is important to be alert to misconceptions such as those presented in the table.

Table 4.18. Common Misconceptions About the Concepts in Lesson 6

Topic	Student Misconception	Explanation
Scaling	"To scale" means to make something larger.	Scale factor used may be greater than 1 if the depiction is larger than the item being represented (e.g., cell representation) or less than 1, such as a bridge.
Variables	There is confusion about which are the independent and dependent variables.	An independent variable is a variable that you can control (span length of a bridge). A dependent variable is a variable that you observe and measure (weight the span will hold).
Models	Models are only in 3-D.	Models in fact can be 2-D representations such as pictures, graphs, written descriptions, blueprints, and other representations, as well as a 3-D model.
	A model's usefulness is based solely on the model's physical resemblance to the object being modeled.	A physical model is a smaller or larger physical copy of an object such as a bridge. The model represents a similar object in the sense that scale is an important characteristic of the model.

PREPARATION FOR LESSON 6

Gather materials for the mathematics investigation and make copies of the student handouts attached at the end of this lesson. Assign each project group one of the six bridge scenarios found on the Final Project Bridge Scenarios handout (p. 204).

Arrange for a member of the local department of transportation to speak to the class about a topic related to bridges or infrastructure in the local community. Possible topics that could be addressed in the presentation include the following:

- The current state of the infrastructure in the local community.
- Current building projects (just completed, in process, or about to begin).
- The process the department uses to select a bridge design.

Invite other teachers and members of the community to hear the students' presentations on the final day of the unit. Identify a panel of judges (e.g., other teachers) to ask questions of the debate teams and determine the winner of the WPA debate in social studies.

LEARNING COMPONENTS

Introductory Activity/Engagement

Connection to the Challenge: Begin each day of this lesson by directing students' attention to the driving question for the module and challenge: How can we develop a decision model to help us make a recommendation to the local department of transportation on the type of bridge to build for a given location? Ask students why bridges are so important and what impact a bridge might have on a community. Hold a brief student discussion of how their learning in the previous days' lesson(s) contributed to their ability to create their plan and build their prototype. You may wish to hold a class discussion, creating a class list of key ideas on chart paper, or you may wish to have students create a notebook entry with this information.

Driving Question for Lesson 6: Given specific environmental conditions, what is the most cost-effective bridge to build?

Mathematics Class: Ask the students to consider the following:

- Think about the bridges that we have studied over the past weeks. In your groups, discuss what characteristics they all have in common.
- What are some characteristics that are unique to the four types of bridges (beam, arch, suspension, cable-stayed)?

Remind students of the module challenge and hand out the Bridge Design Challenge student packet (p. 202). Review the contents of the packet with students and tell them that before they complete the challenge they will consider factors that influence decisions about what type of bridges are built.

Science Connection: Not applicable.

ELA Connection: Engage students in a discussion about a time when they were forced to make a hard decision. As they discuss, help students see that most decisions require us to analyze numerous pieces of information to make the best decision possible.

Social Studies Connection: Not applicable.

Activity/Exploration

Mathematics and Science Class: Say to students, "Now that we have explored several different designs of bridges, we are going to begin to explore the types of sites for which they are best suited." Assign each project group two of the six bridge scenarios that can be found on the Final Project Bridge Scenarios handout (p. 204). Place students in their project groups and give them the Bridge Facts handout (p. 205). For this investigation, students need access to a computer lab, laptops, or tablets to do their research.

In this activity, students work together to gain a more comprehensive understanding of the four types of bridges they have been studying. Students will research span length, function, ideal location, cost, and expected longevity of each bridge design. This information will be used to help them develop their decision model.

Once students have completed their investigation, bring them together as a class and discuss their findings. Ask the following questions:

- What type of bridge did you find that could span the longest distance? (As you discuss this question, it is important to discuss the difference between longest bridge and longest span length in between piers, towers, or supports. For example, the Lake Pontchartrain Bridge is very long, but no single span is very long. Both of these factors are good to consider.)
- What functions did various bridges serve? Are there bridges that are better suited over waterways where there is a lot of traffic on the water?
- What characteristics of the site were important to the various bridge designs?
- What bridge did your group find to be the most expensive to build? To maintain? How does this affect a decision of whether to use this type of bridge?
- None of the bridges that we build last forever. With this in mind, we should have an idea of how long we think they will last. What did you find about the approximate life of the bridge designs? Did you find any factors that either increased or decreased the expected longevity of a bridge?

ELA Connection: Introduce students to the idea of decision models:

- What are they?
- How do they work?

- What are decisions you have to make every day?
- How do you make those decisions?
- Is there a way that we can help ensure that we are making the correct decision?

Social Studies Connection: Students continue to research information on the WPA.

Explanation

Mathematics and Science Class: Bring in a member from the local department of transportation to present to the class. Possible topics that could be addressed in the presentation include the following:

- The current state of the infrastructure in the local community.
- Current building projects (just completed, in process, or about to begin).
- The process the department uses to select a bridge design.

ELA Connection: Students work with the decision models developed in mathematics class to develop a written proposal that could be presented to the local department of transportation to use at the local council meeting.

Social Studies Connection: Divide students into two teams. Pose the following scenario to the class: *One presidential candidate believes that we are in need of another program like the WPA. The candidate believes a program such as this will help improve the infrastructure of the United States and provide jobs to unemployed people in America. The other candidate disagrees and believes that it is not an efficient way to improve the infrastructure and that it is more cost effective to provide other government benefits to the unemployed.*

Assign one of the two stances to each team. Present students with the debate format you will use (see the Teacher Background Information section on p. 195 for a suggestion). Tell the teams that they should each have two or three major points to make their argument and that they should predict what the opposing team's points will be and be prepared to respond to those points. Give students time to develop their arguments and responses.

Elaboration/Application of Knowledge

Mathematics and Science Class: Instruct students to take the information they have collected about the four types of bridges and with their team develop a decision model that would help them determine the appropriate bridge design given a particular scenario. Once each team of students has developed their decision model, they can test it using the following website: *www.pbs.org/wgbh/nova/tech/build-bridge-p1.html.*

Provide students with a copy of the Final Project Overview handout (p. 203) and a description of their proposed bridge site. Students will orally present their final projects to classmates and members of the community using their posters and written papers.

ELA Connection: Conduct a peer critique of written proposals.

Social Studies Connection: Conduct the debate. Have a panel of "judges" available to ask questions of the teams and declare a winner.

Evaluation/Assessment

Students may be assessed on the following performance tasks and other measures listed.

Performance Tasks

- Student teams present their proposals to the class and potentially members of the local department of transportation and other community members, too. The students present their decision models and their recommendations for the appropriate bridge decision for their assigned locations in a written proposal and poster. Rubrics for the written proposal and poster and presentation are attached at the end of this lesson.
- Students debate the need for another program like the previous WPA. A rubric for the debate is attached at the end of this lesson.

Other Measures

- At the conclusion of the lesson, have students reflect in their STEM Research Notebooks about their current understanding of bridge design, structure, and function.
- Have students analyze how their understanding has changed. Students could do this in an essay or a class discussion.
- Assess students on their collaboration during their final group project using the Collaboration Rubric (attached at the end of this lesson).
- Finally, close the lesson by having students reflect in their STEM Research Notebooks about all they have learned and what they still would like to learn more about.

INTERNET RESOURCES

Information about decision models

- *www.web-books.com/eLibrary/ON/B0/B58/070MB58.html*
- *www.mckinsey.com/insights/strategy/the_benefits_and_limits_of_decision_models*
- *www.business-analysis-made-easy.com/Decision-Making-Models.html*
- *www.ncbi.nlm.nih.gov/pmc/articles/PMC2245736*
- *www.mindtools.com/pages/article/newTED_91.htm*

"Build a Bridge" article

- *www.pbs.org/wgbh/nova/tech/build-bridge-p1.html*

STUDENT PACKET, PAGE 1

BRIDGE DESIGN CHALLENGE

STUDENT PACKET CONTENTS

✓ Final Project Overview

✓ Final Project Bridge Scenarios handout

✓ Bridge Facts handout

✓ Assessment

- Written Proposal Rubric
- Poster and Presentation Rubric
- Collaboration Rubric

STUDENT PACKET, PAGE 2

BRIDGE DESIGN CHALLENGE

FINAL PROJECT OVERVIEW

Over the past weeks, we have discussed, investigated, and researched the design, structure, and function of four different types of bridges. It has become apparent that not every bridge design is a good fit for every site. For this project, you will be given a description of the location for a proposed site for a bridge with some details about the necessary requirements of the bridge that will be constructed.

REQUIREMENTS

- ✓ Include a copy of the decision model your group has created and tested.
- ✓ Apply your decision model to the scenario you have been given to determine the most appropriate bridge design.
- ✓ Develop a written proposal that could be presented to the local department of transportation explaining and defending your decision.
- ✓ Create a poster displaying your decision model, the description of bridge site, the bridge design you selected, and the rationale for why you believe it is the correct design.
- ✓ Orally present your poster to your peers, members of the local department of transportation, and other members of the community.

Your project will be evaluated using the rubrics attached (Written Proposal Rubric, Poster and Presentation Rubric, and Collaboration Rubric).

STUDENT PACKET, PAGE 3

BRIDGE DESIGN CHALLENGE

FINAL PROJECT BRIDGE SCENARIOS

Scenario 1

This location requires a multilane bridge that must be able to support heavy traffic loads. The bridge must span a distance of 1,900 feet over a waterway that is commonly used by sailboats, large yachts, and other recreational boats. Because this bridge will be constructed in a busy tourism area, city planners have requested a bridge that is less common and visually appealing.

Scenario 2

To improve the health of the community, city planners are constructing a jogging trail that will run through one of the city parks. This location requires a small bridge (125 feet) that will be used mostly by pedestrians. Since this is a small project, the city would like to spend as little as possible.

Scenario 3

As industry production has increased, so has railroad traffic. To help accommodate this influx of rail traffic, a new railroad is being constructed. The bridge in this location must span a length of 450 feet and be strong enough to support heavy railroad cars. Further, because this bridge must be constructed over a large, deep, and rocky canyon, this bridge must be able to withstand high winds.

Scenario 4

This location requires a large bridge that will extend a new highway being constructed. The bridge must span 4,500 feet over a large body of water that is constantly used by commercial shipping companies. Because of the frequency of commercial shipping traffic, it is important that this structure not disrupt the flow of traffic on the water.

Scenario 5

This location requires a bridge that is approximately five miles long. This bridge is being constructed to connect two large metropolitan areas that are on opposite sides of a large lake. While this bridge must be able to support a great deal of commuter traffic, there is no need to provide passage for large boats.

Scenario 6

It has been determined that a highway needs to pass through a national park. However, a road would split the park and prevent wildlife from moving freely. The bridge must be 1.5 miles in total length, but have as little impact as possible; therefore, it is important to minimize the number of towers used. Because of the high initial and maintenance costs associated with suspension bridges, city planners have asked that a suspension bridge not be used.

STUDENT PACKET, PAGE 4

BRIDGE DESIGN CHALLENGE

BRIDGE FACTS

Over the past weeks, we have explored the strength and the effects of span length on four different types of bridges. In your groups, take another look at these bridges and begin to think about what makes them the appropriate choice for a given site. For each of the four bridges we have discussed (beam, arch, suspension, and cable-stayed), you and your group will gather information to help gain a more comprehensive understanding of these bridges.

Span Length

For this section, you and your group will take a look at the ideal span length for each type of bridge. Are the span lengths normally relatively short? What are the limits to how far each type of bridge can span? Can you find the longest example of each type of bridge?

Function

In this section, your group will explore the function of each type of bridge. Where is each type of bridge typically found? Are they found over waterways? Are they constructed to hold extremely heavy loads? Are they designed to be both functional as well as attractive? Are there any special characteristics that we did not discuss in class?

Ideal Location

What is true about the geography of where each bridge is found? What are characteristics that must be present for construction of each type of bridge to be feasible?

Cost

In this section, you are to look at the cost of each type of bridge. Try to find an example of each type of bridge and how much it costs to build and how much it costs to maintain. Which bridge design seems to be the most expensive to build? Which bridge design will be the cheapest?

Longevity

As we discussed earlier in the unit, most bridges in this country are old and decaying. How long do we expect each of the designs to last? Are there some that are expected to last longer than others?

Other

Is there something you came across in your research that might be important in determining whether a particular type of bridge would be a good fit for a given location?

Written Proposal Rubric

Name: ______________________________

Criteria	Below Standard (1 point)	Approaching Standard (2 points)	Meets Standard (3 points)	Exceeds Standard (4 points)	*Score*
ORGANIZATION	Information is unorganized and inaccurate.	Information is a little disorganized and paragraphs are not well-constructed.	Information is organized with well-constructed paragraphs and information is factual and correct.	Information is very organized with well-constructed paragraphs and subheadings; information is factual and correct.	
QUALITY OF INFORMATION	Information has little or nothing to do with the main topic.	Information clearly relates to the decision model, but no details or supporting examples were used.	Information clearly provides a rationale for the decision model; it includes one or two supporting details or examples.	Information clearly provides a rationale for the decision model; it includes several supporting details or examples.	
VISUAL	No decision tree or decision table was provided.	Lacks either a decision table or a decision tree or these do not reflect the decision model.	A complete decision table and decision tree are included in the paper that directly reflect the decision model.	A well-developed decision table and decision tree are included in the paper that directly reflect the decision model.	
MECHANICS	Proposal has too many grammatical, spelling, or punctuation errors.	Proposal has many grammatical, spelling, or punctuation errors.	The proposal has almost no grammatical, spelling, or punctuation errors.	There are no grammatical, spelling, or punctuation errors in the proposal.	
SOURCES	Too many sources are not documented accurately or it is evident that they are not documented at all.	Uses a couple of types of sources but does not always document the use of these throughout the paper or in the reference page.	Uses a couple of types of sources and documents the use of these throughout the paper and in the reference page.	Uses a variety of sources and documents the use of these throughout the paper and in the reference page.	

TOTAL SCORE: ________

COMMENTS:

Poster and Presentation Rubric Name: ____				
Criteria	Below Standard	Approaching Standard	Meets or Exceeds Standard	*Comments*
EXPLANATION OF DECISION MODEL	The poster either did not include an explanation or the explanation was incomplete.	The poster provided an explanation but the explanation of the group's decision model was confusing or difficult to follow.	The poster provided a clear explanation of the group's decision model. The information included was well thought out and concise.	
RESPONSE TO AUDIENCE QUESTIONS	The students did not respond to the audience's questions and feedback. They were unprepared to answer questions.	The students were able to answer some of the audience's questions but not all. If students didn't know an answer, they stated that they didn't know or tried to make up an answer.	The students responded to audience's questions and feedback. When students didn't understand an audience member's question, they asked for clarification. If the student didn't know the answer to a question, they admitted that they did not know the answer and explained how they would find the answer.	
PARTICIPATION IN GROUP PROJECT AND PRESENTATION	The student minimally participated in the development of the presentation and decision model and had a minimal role in the presentation.	The student participated in the development of the presentation and decision model but had a minimal role in the presentation.	The student participated in the development of the presentation and decision model and actively participated in the presentation.	

Collaboration Rubric

Name: ____________________

Criteria	Below Standard (1 point)	Approaching Standard (2 points)	Meets Standard (3 points)	Exceeds Standard (4 points)	*Score*
CONTRIBUTIONS/ PARTICIPATION ATTITUDE	Seldom cooperative, rarely offers useful ideas. Is disruptive.	Sometimes cooperative, sometimes offered useful ideas. Rarely displays positive attitude.	Cooperative, usually offered useful ideas. Generally displays positive attitude.	Always willing to help and do more, routinely offered useful ideas. Always displays positive attitude.	
WORKING WITH OTHERS/ COOPERATION	Did not do any work—does not contribute, does not work well with others, usually argues with teammates.	Could have done more of the work—has difficulty. Requires structure, directions, and leadership, sometimes argues.	Did his or her part of the work—cooperative. Works well with others, rarely argues.	Did more than others—highly productive. Works extremely well with others, never argues.	
FOCUS ON TASK/ COMMITMENT	Often is not a good team member. Does not focus on the task and what needs to be done. Lets others do the work.	Sometimes not a good team member. Sometimes focuses on the task and what needs to be done. Must be prodded and reminded to keep on task.	Does not cause problems in the group. Focuses on the task and what needs to be done most of the time. Can count on this person.	Tries to keep people working together. Almost always focused on the task and what needs to be done. Is very self-directed.	
COMMUNICATION/ LISTENING INFORMATION SHARING	Rarely listens to, shares with, or supports the efforts of others. Is always talking and never listens to others. Provided no feedback to others. Does not relay any information to teammates.	Often listens to, shares with, and supports the efforts of others. Usually does most of the talking—rarely listens to others. Provided little feedback to others. Relays very little information—some relates to the topic.	Usually listens to, shares with, and supports the efforts of others. Sometimes talks too much. Provided some effective feedback to others. Relays some basic information—most relates to the topic.	Always listens to, shares with, and supports the efforts of others. Provided effective feedback to other members. Relays a great deal of information—all relates to the topic.	

TOTAL SCORE: ________

COMMENTS:

Social Studies Debate Rubric

Name: ____________

Criteria	Below Standard (1 point)	Approaching Standard (2 points)	Meets Standard (3 points)	Exceeds Standard (4 points)	*Score*
POSITION/STANCE	The students did not present a side to the argument. Their position was solely based on opinion. No support was provided.	The students did not clearly present their side to the argument. They presented some facts but most of the argument was based on opinion.	The students presented their position clearly and provided factual information to support their argument. However, they did not cite their sources.	The students presented their position clearly and provided factual information to support their argument and cited their sources.	
UNDERSTANDING OF WPA TRANSLATION INTO TODAY'S SOCIETY	The students displayed little to no understanding of the WPA and its impact on society in the 1920s and 1930s. It is clear that the students did not research the topic in depth.	The students demonstrated a surface-level understanding of the WPA but did not demonstrate a thorough understanding of the topic. The students' connection to today's landscape was absent.	The students demonstrated a good understanding of the impact of the WPA during the 1920s and 1930s but only provided some information on how this could or not be effective in today's landscape.	The students demonstrated a deep understanding of the impact of the WPA during the 1920s and 1930s and discussed clearly how this could or could not be effective in today's landscape.	
PARTICIPATION IN DEBATE	The student did not participate in the research or in the debate.	The student participated in the research of the topic but did not come prepared to debate the topic. The student did not participate in the debate.	The student participated in the research of the topic and had a small part in the debate.	The student participated in the research of the topic and the debate. The student provided strong evidence to refute the claims of the opposing side.	
STUDENT VOICE	The student's voice was not heard in the debate OR the student made comments that were inappropriate, mean, or hateful.	The student used a strong voice but was not clear and confident. At times, the student tried to engage with the audience.	The student used a clear, confident, loud, and expressive voice. However, the student did not make eye contact with the audience and didn't always engage with the audience in a respectful tone.	The student used a clear, confident, loud, and expressive voice. The student made eye contact with the audience and engaged with the audience in a respectful tone.	

TOTAL SCORE: ________

COMMENTS:

ADDITIONAL RESOURCES

Beach, J. 2001. *The three-mile bridge: Across Pensacola Bay on a span of poems.* Bloomington, IN: AuthorHouse.

Floca, B. 2013. *Locomotive.* New York: Atheneum Books for Young Readers.

Lundberg, J. 2009. The poetry of the Brooklyn Bridge. *www.huffingtonpost.com/john-lundberg/the-poetry-of-the-brookly_b_211935.html.*

Mackay, D. A. 2010. *The Building of Manhattan.* New York: Holt McDougal.

Michigan Department of Transportation. 2015. From plans to pavement: How a road is built. *www.michigan.gov/mdot/0,1607,7-151-9615-129011--,00.html.*

Paty, A. H. 2000. Rocks and minerals—foundations of society. *Science Scope* 23 (8): 30–31.

PBS Learning Media. 2006. "Rocks and Minerals" video. *www.pbslearningmedia.org/resource/idptv11.sci.ess.earthsys.d4krom/rocks-and-minerals.*

PBS Learning Media. 2006. USGS: Minerals in Our Environment. *www.pbslearningmedia.org/resource/ess05.sci.ess.earthsys.mineralenv/minerals-in-our-environment.*

Rossi, D. W. 2004. Using elementary interactive science journals to encourage reflection, learning and positive attitudes toward science. *www.utdallas.edu/sme/files/Using_Elementary_Interactive_Science_JournalsDRW.pdf.*

SmartR Virtual Learning Experiences. 2012. "Science: Rocks" video. *http://smartr.edc.org/sciencerocks.*

Weathers, J., J. Galloway, and D. Frank. 2001. USGS: Minerals in our environment. *http://pubs.usgs.gov/of/2000/0144.* (Printable poster of minerals in our home environments)

Williams, J. S., ed. 2013. Motionless from the iron bridge: A northwest anthology of bridge poems. Available from the Oregon Poetry Association. *http://oregonpoets.org/motionless-from-the-iron-bridge-a-northwest-anthology-of-bridge-poems.*

Yezerski, T. F. 2011. *Meadowlands: A wetlands survival story.* New York: Farrar, Straus & Giroux.

REFERENCES

Alchin, L. 2015. Roman arches. *www.tribunesandtriumphs.org/roman-architecture/roman-arches.htm.*

Eduplace. 2015. Writing about history: Problem and solution. Boston: Houghton Mifflin Company. *www.eduplace.com/kids/socsci/ca/books/bkf3/writing/06_romarch.pdf.*

Harvey, S., and A. Goudvis. 2007. *Strategies that work: Teaching comprehension for understanding and engagement.* 2nd ed. Portland, ME: Stenhouse.

Oxford Dictionaries. 2015. Infrastructure definition. *www.oxforddictionaries.com/us/definition/american_english/infrastructure.*

Public Broadcasting Service (PBS). 2001. Building big: Glossary of engineering words. *www.pbs.org/wgbh/buildingbig/glossary_head.html.*

Roth, S. L., and C. Trumbore. 2011. *The mangrove tree: Planting trees to feed families*. New York: Lee & Low Books.

Sanders, J., and J. Moudy. 2008. Literature apprentices: Understanding nonfiction text structures with mentor texts. *Journal of Children's Literature* 34 (2): 31–40.

Short, K. G., and J. C. Harste. 1996. *Creating classrooms for authors and inquirers*. 2nd ed. Portsmouth, NH: Heinemann.

Simmons, B. 2014. Mathwords scale factor definition. *www.mathwords.com/s/scale_factor.htm*.

5

TRANSFORMING LEARNING WITH IMPROVING BRIDGE DESIGN AND THE *STEM ROAD MAP CURRICULUM SERIES*

Carla C. Johnson

This chapter serves as a conclusion to the Improving Bridge Design integrated STEM curriculum module, but it is just the beginning of the transformation of your classroom that is possible through use of the *STEM Road Map Curriculum Series*. In this book, many key resources have been provided to make learning meaningful for your students through integration of science, technology, engineering, and mathematics, as well as social studies and English language arts, into powerful problem- and project-based instruction. First, the Improving Bridge Design curriculum is grounded in the latest theory of learning for students in grade 8 specifically. Second, as your students work through this module, they engage in using the engineering design process (EDP) and build prototypes like engineers and STEM professionals in the real world. Third, students acquire important knowledge and skills grounded in national academic standards in mathematics, English language arts, science, and 21st century skills that will enable their learning to be deeper, retained longer, and applied throughout, illustrating the critical connections within and across disciplines. Finally, authentic formative assessments, including strategies for differentiation and addressing misconceptions, are embedded within the curriculum activities.

The Improving Bridge Design curriculum in The Represented World STEM Road Map theme can be used in single-content classrooms (e.g., mathematics) where there is only one teacher or expanded to include multiple teachers and content areas across classrooms. Through the exploration of the Bridge Design Challenge, students engage in a real-world STEM problem on the first day of instruction and gather necessary knowledge and skills along the way in the context of solving the problem.

The other topics in the STEM Road Map Curriculum Series are designed in a similar manner, and NSTA Press has additional volumes in this series for this and other grade levels and plans to publish more. The volumes covering Innovation and Progress have been published and are as follows:

- *Amusement Park of the Future, Grade 6*
- *Construction Materials, Grade 11*
- *Harnessing Solar Energy, Grade 4*
- *Transportation in the Future, Grade 3*
- *Wind Energy, Grade 5*

The tentative list of other books includes the following themes and subjects:

- The Represented World
  - Car crashes
  - Changes over time
  - Packaging design
  - Patterns and the plant world
  - Radioactivity
  - Rainwater analysis
  - Swing set makeover
- Cause and Effect
  - Influence of waves
  - Hazards and the changing environment
  - The role of physics in motion
- Sustainable Systems
  - Creating global bonds
  - Composting: Reduce, reuse, recycle
  - Hydropower efficiency
  - System interactions

- Optimizing the Human Experience
    - Genetically modified organisms
    - Mineral resources
    - Rebuilding the natural environment
    - Water conservation: Think global, act local

If you are interested in professional development opportunities focused on the STEM Road Map specifically or integrated STEM or STEM programs and schools overall, contact the lead editor of this project, Dr. Carla C. Johnson (*carlacjohnson@purdue.edu*), associate dean and professor of science education at Purdue University. Someone from the team will be in touch to design a program that will meet your individual, school, or district needs.

APPENDIX

CONTENT STANDARDS ADDRESSED IN THIS MODULE

NEXT GENERATION SCIENCE STANDARDS

Table A1 (p. 218) lists the science and engineering practices, disciplinary core ideas, and crosscutting concepts this module addresses. The supported performance expectations are as follows:

- MS-ESS2-1. Develop a model to describe the cycling of Earth's materials and the flow of energy that drives this process.
- MS-ESS3-1. Construct a scientific explanation based on evidence for how the uneven distributions of Earth's mineral, energy, and groundwater resources are the result of past and current geoscience processes.
- MS-ETS1-4. Develop a model to generate data for iterative testing and modification of a proposed object, tool, or process such that an optimal design can be achieved.

Table A1. *Next Generation Science Standards (NGSS)*

Science and Engineering Practices
DEVELOPING AND USING MODELS Modeling in 6–8 builds on K–5 experiences and progresses to developing, using, and revising models to describe, test, and predict more abstract phenomena and design systems. • Develop and use a model to describe phenomena. • Develop a model to generate data to test ideas about designed systems, including those representing inputs and outputs.
ANALYZING AND INTERPRETING DATA Analyzing data in 6–8 builds on K–5 experiences and progresses to extending quantitative analysis to investigations, distinguishing between correlation and causation, and basic statistical techniques of data and error analysis. • Analyze and interpret data to provide evidence for phenomena.
CONSTRUCTING EXPLANATIONS AND DESIGNING SOLUTIONS Constructing explanations and designing solutions in 6–8 builds on K–5 experiences and progresses to include constructing explanations and designing solutions supported by multiple sources of evidence consistent with scientific ideas, principles, and theories. • Construct a scientific explanation based on valid and reliable evidence obtained from sources (including the students' own experiments) and the assumption that theories and laws that describe the natural world operate today as they did in the past and will continue to do so in the future.
SCIENTIFIC KNOWLEDGE IS OPEN TO REVISION IN LIGHT OF NEW EVIDENCE (NATURE OF SCIENCE PRACTICE) • Science findings are frequently revised and/or reinterpreted based on new evidence.
Disciplinary Core Ideas
ESS2.A: EARTH'S MATERIALS AND SYSTEMS • All Earth processes are the result of energy flowing and matter cycling within and among the planet's systems. This energy is derived from the sun and Earth's hot interior. The energy that flows and matter that cycles produce chemical and physical changes in Earth's materials and living organisms.
ESS3.A: NATURAL RESOURCES • Humans depend on Earth's land, ocean, atmosphere, and biosphere for many different resources. Minerals, fresh water, and biosphere resources are limited, and many are not renewable or replaceable over human lifetimes. These resources are distributed unevenly around the planet as a result of past geologic processes.

Continued

Table A1. *(continued)*

Disciplinary Core Ideas *(continued)*
ETS1.B: DEVELOPING POSSIBLE SOLUTIONS • A solution needs to be tested, and then modified on the basis of the test results, in order to improve it. • Models of all kinds are important for testing solutions.
ETS1.C: OPTIMIZING THE DESIGN SOLUTION • The iterative process of testing the most promising solutions and modifying what is proposed on the basis of the test results leads to greater refinement and ultimately to an optimal solution.
Crosscutting Concepts
STABILITY AND CHANGE • Explanations of stability and change in natural or designed systems can be constructed by examining the changes over time and processes at different scales, including the atomic scale.
PATTERNS • Patterns in rates of change and other numerical relationships can provide information about natural systems.
CAUSE AND EFFECT • Cause and effect relationships may be used to predict phenomena in natural or designed systems.
INFLUENCE OF SCIENCE, ENGINEERING, AND TECHNOLOGY ON SOCIETY AND THE NATURAL WORLD • All human activity draws on natural resources and has both short- and long-term consequences, positive as well as negative, for the health of people and the natural environment.

Source: NGSS Lead States. 2013. *Next Generation Science Standards: For states, by states.* Washington, DC: National Academies Press. *www.nextgenscience.org/next-generation-science-standards.*

Table A2. Common Core Mathematics and English Language Arts (ELA) Standards

MATHEMATICAL PRACTICES	READING STANDARDS
• MP1. Make sense of problems and persevere in solving them. • MP2. Reason abstractly and quantitatively. • MP3. Construct viable arguments and critique the reasoning of others. • MP4. Model with mathematics. • MP5. Use appropriate tools strategically. • MP6. Attend to precision. • MP7. Look for and make use of structure. • MP8. Look for and express regularity in repeated reasoning. MATHEMATICAL CONTENT • 8.EE.B.5. Graph proportional relationships, interpreting the unit rate as the slope of the graph. Compare two different proportional relationships represented in different ways. For example, compare a distance-time graph to a distance-time equation to determine which of two moving objects has greater speed. • 8.EE.C.7.B. Solve linear equations with rational number coefficients, including equations whose solutions require expanding expressions using the distributive property and collecting like terms.	• RL.8.1. Cite the textual evidence that most strongly supports an analysis of what the text says explicitly as well as inferences drawn from the text. • RL.8.2. Determine a theme or central idea of a text and analyze its development over the course of the text, including its relationship to the characters, setting, and plot; provide an objective summary of the text. • RI.8.7. Evaluate the advantages and disadvantages of using different mediums (e.g., print or digital text, video, multimedia) to present a particular topic or idea. WRITING STANDARDS • W.8.1. Write arguments to support claims with clear reasons and relevant evidence. • W.8.1.A. Introduce claim(s), acknowledge and distinguish the claim(s) from alternate or opposing claims, and organize the reasons and evidence logically. • W.8.1.B. Support claim(s) with logical reasoning and relevant evidence, using accurate, credible sources and demonstrating an understanding of the topic or text. • W.8.1.C. Use words, phrases, and clauses to create cohesion and clarify the relationships among claim(s), counterclaims, reasons, and evidence. • W.8.1.E. Provide a concluding statement or section that follows from and supports the argument presented. • W.8.2. Write informative/explanatory texts to examine a topic and convey ideas, concepts, and information through the selection, organization, and analysis of relevant content.

Continued

Table A2. *(continued)*

MATHEMATICAL CONTENT	WRITING STANDARDS
(continued) • 8.F.B.5. Describe qualitatively the functional relationship between two quantities by analyzing a graph (e.g., where the function is increasing or decreasing, linear or nonlinear). Sketch a graph that exhibits the qualitative features of a function that has been described verbally. • 8.G.B.7. Apply the Pythagorean theorem to determine unknown side lengths in right triangles in real-world and mathematical problems in two and three dimensions.	*(continued)* • W.8.2.A. Introduce a topic clearly, previewing what is to follow; organize ideas, concepts, and information into broader categories; include formatting (e.g., headings), graphics (e.g., charts, tables), and multimedia when useful to aiding comprehension. • W.8.2.B. Develop the topic with relevant, well-chosen facts, definitions, concrete details, quotations, or other information and examples. • W.8.2.C. Use appropriate and varied transitions to create cohesion and clarify the relationships among ideas and concepts. • W.8.2.D. Use precise language and domain-specific vocabulary to inform about or explain the topic. • W.8.2.E. Establish and maintain a formal style. • W.8.2.F. Provide a concluding statement or section that follows from and supports the information or explanation presented. • W.8.3. Write narratives to develop real or imagined experiences or events using effective technique, relevant descriptive details, and well-structured event sequences. • W.8.3.A. Engage and orient the reader by establishing a context and point of view and introducing a narrator and/or characters; organize an event sequence that unfolds naturally and logically. • W.8.3.D. Use precise words and phrases, relevant descriptive details, and sensory language to capture the action and convey experiences and events. • W.8.6. Use technology, including the internet, to produce and publish writing and present the relationships between information and ideas efficiently as well as to interact and collaborate with others. • W.8.7. Conduct short research projects to answer a question (including a self-generated question), drawing on several sources and generating additional related, focused questions that allow for multiple avenues of exploration.

Continued

Table A2. *(continued)*

	WRITING STANDARDS *(continued)* • W.8.8. Gather relevant information from multiple print and digital sources, using search terms effectively; assess the credibility and accuracy of each source; and quote or paraphrase the data and conclusions of others while avoiding plagiarism and following a standard format for citation. **SPEAKING AND LISTENING STANDARDS** • SL.8.1. Engage effectively in a range of collaborative discussions (one-on-one, in groups, and teacher-led) with diverse partners on grade 8 topics, texts, and issues, building on others' ideas and expressing their own clearly. • SL.8.1.A. Come to discussions prepared, having read or researched material under study; explicitly draw on that preparation by referring to evidence on the topic, text, or issue to probe and reflect on ideas under discussion. • SL.8.1.B. Follow rules for collegial discussions and decision making, track progress toward specific goals and deadlines, and define individual roles as needed. • SL.8.1.C. Pose questions that connect the ideas of several speakers and respond to others' questions and comments with relevant evidence, observations, and ideas. • SL.8.1.D. Acknowledge new information expressed by others, and, when warranted, qualify or justify their own views in light of the evidence presented. • SL.8.2. Analyze the purpose of information presented in diverse media and formats (e.g., visually, quantitatively, orally) and evaluate the motives (e.g., social, commercial, political) behind its presentation. • SL.8.3. Delineate a speaker's argument and specific claims, evaluating the soundness of the reasoning and relevance and sufficiency of the evidence and identifying when irrelevant evidence is introduced.

Continued

Table A2. *(continued)*

	SPEAKING AND LISTENING STANDARDS (*continued*) • SL.8.4. Present claims and findings, emphasizing salient points in a focused, coherent manner with relevant evidence, sound valid reasoning, and well-chosen details; use appropriate eye contact, adequate volume, and clear pronunciation. • SL.8.5. Integrate multimedia and visual displays into presentations to clarify information, strengthen claims and evidence, and add interest. • SL.8.6. Adapt speech to a variety of contexts and tasks, demonstrating command of formal English when indicated or appropriate. LANGUAGE STANDARDS • L.8.5. Demonstrate understanding of figurative language, word relationships, and nuances in word meanings. • L.8.5.A. Interpret figures of speech (e.g., verbal irony, puns) in context. • L.8.5.B. Use the relationship between particular words to better understand each of the words.

Source: National Governors Association Center for Best Practices and Council of Chief State School Officers (NGAC and CCSSO). 2010. *Common core state standards.* Washington, DC: NGAC and CCSSO.

Table A3. 21st Century Skills From the Framework for 21st Century Learning

21st Century Skills	Learning Skills and Technology Tools	Teaching Strategies	Evidence of Success
INTERDISCIPLINARY THEMES	• Global Awareness • Environmental Literacy • Financial, Economic, Business, and Entrepreneurial Literacy • Civic Literacy	• Discuss U.S. infrastructure (e.g., roads, bridges) and analyze recent bridge collapses • Investigate the impact that bridges have had on individuals and local culture • Identify how earth materials are used in everyday life.	• Creation of multimedia presentation on bridge uses in the state • Mathematics STEM Research Notebook entries • Science STEM Research Notebook entries
LEARNING AND INNOVATION SKILLS	• Creativity and Innovation • Critical Thinking and Problem Solving • Communication and Collaboration	• Analyze data and interpret trends in order to develop a decision model • Observation of rock samples • Investigation of ice wedging on roads	• Written proposal and poster presentation • Rock Observation Assessment Rubric • Science STEM Research Notebook reflections
INFORMATION, MEDIA, AND TECHNOLOGY SKILLS	• Information Literacy • Media Literacy • ICT Literacy	• Applying research skills to (1) determine common minerals in the state and (2) explore bridge design, structure, and function.	• Creation of multimedia presentation on mineral uses in the state. • Written proposal and poster presentation
LIFE AND CAREER SKILLS	• Flexibility and Adaptability • Initiative and Self-Direction • Social and Cross-Cultural Skills • Productivity and Accountability • Leadership and Responsibility	• Establish collaborative learning expectations • Provide guidelines for effective peer critique and how to use this feedback for future projects	• Collaboration Rubric • Rock Observation Assessment Rubric

Source: Partnership for 21st Century Learning. 2015. Framework for 21st Century Learning. *www.p21.org/our-work/p21-framework.*

Table A4. English Language Development (ELD) Standards

ELD STANDARD 1: SOCIAL AND INSTRUCTIONAL LANGUAGE
English language learners communicate for Social and Instructional purposes within the school setting.
ELD STANDARD 2: THE LANGUAGE OF LANGUAGE ARTS
English language learners communicate information, ideas, and concepts necessary for academic success in the content area of Language Arts.
ELD STANDARD 3: THE LANGUAGE OF MATHEMATICS
English language learners communicate information, ideas, and concepts necessary for academic success in the content area of Mathematics.
ELD STANDARD 4: THE LANGUAGE OF SCIENCE
English language learners communicate information, ideas, and concepts necessary for academic success in the content area of Science.
ELD STANDARD 5: THE LANGUAGE OF SOCIAL STUDIES
English language learners communicate information, ideas, and concepts necessary for academic success in the content area of Social Studies.
COMPLEMENTARY STRAND: THE LANGUAGE OF TECHNOLOGY AND ENGINEERING
Students at all levels of English language proficiency interact with grade-level words and expressions, such as *software program, file name, tool bar, icons, formatting, image, clip art, slides,* and *multimedia presentation.*

Source: WIDA. 2012. 2012 amplification of the English language development standards: Kindergarten–grade 12, *www.wida.us/standards/eld.aspx.*

INDEX

Page numbers printed in **boldface type** indicate tables, figures, or handouts.